DESERT EDENS

HISTORIES OF ECONOMIC LIFE

Jeremy Adelman, Sunil Amrith,
Emma Rothschild, and Francesca Trivellato, Series Editors

Desert Edens

COLONIAL CLIMATE ENGINEERING
IN THE AGE OF ANXIETY

PHILIPP LEHMANN

PRINCETON UNIVERSITY PRESS
PRINCETON & OXFORD

Published by Princeton University Press
41 William Street, Princeton, New Jersey 08540
99 Banbury Road, Oxford OX2 6JX

press.princeton.edu

First paperback printing, 2024
Paper ISBN 978-0-691-23934-7
Cloth ISBN 978-0-691-16886-9
ISBN (e-book) 978-0-691-23828-9

British Library Cataloging-in-Publication Data is available

Editorial: Priya Nelson and Barbara Shi
Production Editorial: Jenny Wolkowicki
Jacket/Cover design: Kimberly Castañeda
Production: Lauren Reese
Publicity: Matthew Taylor and Charlotte Coyne
Copyeditor: Joseph Dahm

Jacket/Cover art: Panropa Tower illustration by Peter Behrens, 1931

This book has been composed in Arno Pro

Yet, he said, it is often our mightiest projects that most obviously betray the degree of our insecurity.

—W. G. SEBALD, *AUSTERLITZ*

CONTENTS

ACKNOWLEDGMENTS

FOR BETTER or worse, history remains a single-author discipline, but that disguises the village needed to research, write, and revise. This book would have been impossible without the encouragement, assistance, and criticism of many individuals who made the process of writing a monograph not only possible, but also highly enjoyable. First and foremost, I am grateful to David Blackbourn, Alison Frank Johnson, Harriet Ritvo, and Emma Rothschild, whose guidance and unwavering support have been invaluable throughout my time working on various iterations of the manuscript. I also owe a great debt to Sunil Amrith, Lydia Barnett, Lino Camprubí, Mark Carey, Lorraine Daston, Alejandra Dubcovsky, Alexander Gall, Matthias Heymann, Fredrik Albritton Jonsson, Charles Maier, Ian Miller, William O'Reilly, Libby Robin, Amberle Sherman, Victor Seow, Josh Specht, and Audra J. Wolfe, who provided comments, advice, and inspiration at different stages of the project. The same goes for the many co-panelists and audience members who engaged with my work at countless workshops and conferences on four different continents. I also extend my gratitude to the anonymous reviewers of the manuscript, whose notes and suggestions were as indispensable and important as they were challenging. Finally, I thank the editors at Princeton University Press, and especially Quinn Fusting and Priya Nelson, whose enthusiasm for the project made it become a book.

The extensive archival research necessary for this project was made possible by generous funding from the CLIR Mellon Fellowship for Dissertation Research in Original Sources and the Deutscher Akademischer Austauschdienst (DAAD). The Center for History and Economics and the Center for European Studies at Harvard University provided additional funding, while the American Council of Learned Societies, the Mellon Foundation, the Hellman Fellows Program, and the Hamburg Institute for Advanced Study provided me with much-needed time and space to work on various rounds of revisions.

My work would have been impossible without the help, knowledge, and patience of the many librarians and archivists whom I encountered during my research. In particular, I would like to thank the staff at Harvard's Widener Library, the library of the Max Planck Institute for the History of Science, the

Tomás Rivera Library at UCR, the Bundesarchiv in Berlin, the archives of the Deutsches Museum, the Staatsbibliothek Berlin, the Bayerische Staatsbibliothek, the Geheimes Staatsarchiv in Berlin, the Niedersächsisches Landesarchiv, the Archives Nationales d'Outre-Mer, the archive of the Académie des Sciences in Paris, and the Archives Diplomatiques in La Courneuve.

I would also like to express my gratitude to my friends, who have stood by my side with "Rat und Tat" and have provided both indispensable advice and welcome distractions during periods of slow progress. I am especially indebted to Scott Phelps, who has been an inspiring force for many years, encouraging me to think more deeply about the larger implications of my work. Lastly, I would like to thank my family. My parents Alena and Frank Lehmann have supported my academic career with great endurance and enthusiasm. My New Mexican "parents" Ingrid and Robert Upton taught me not only English, but also a love for deserts. And my wife Eva Bitran has read through so many drafts of this book that she may know it better than I do. She has endured my bad moods, encouraged me when I needed it most, and celebrated each milestone, no matter how small. Over the past three years my two sons, Elias Leander and Lino Antonin, have joined my supporters' section, but are probably a little disappointed that there are not enough pictures in this book.

Finally, I would like to dedicate this work to the late Maya Peterson, whose intellectual acuity and enthusiasm motivated me every step on the way and whose tragic passing came as an utter shock. I will continue to miss her as both a remarkable colleague and a dear friend.

Introduction

CLIMATE CHANGE AND CHANGING CLIMATES

IN EARLY JULY 1850 the German Sahara explorer Heinrich Barth undertook the arduous task of scaling the southern high plateau of the Libyan Desert. The expedition party had just finished pitching their tents in a particularly dry and barren camping spot when the North African guides pointed out a number of rock engravings to their European travel companions. Barth was exhilarated. On the spot, he drew the images portraying cows, gazelles, and hunting scenes and asked his guides to show him more of the artifacts. Barth also described the engravings in his travel journal, remarking on their possible provenance and their ethnographic and historical significance.[1]

The disconnect between the bodily experience of an arid, denuded desert and the evidence of past human habitation inspired Barth to spend a few lines on the images as "evidence of entirely different living conditions than we are currently observing in these lands." He wrote this without any indication of surprise at the drastic environmental and climatic changes he alluded to. The idea was not new, after all. Offering easily available sources on environmental conditions from antiquity to the present day, North Africa and the Mediterranean region had long been among the preferred sites for Europeans pondering potential climatic shifts. Had Barth still been alive by the last third of the nineteenth century, he might nevertheless have been surprised to witness the fate of his own brief reflections on the subject: his sketches of the Libyan engravings and his cursory notes about "different living conditions" in North Africa's past would become essential material for climatologists engaged in an extended debate on large-scale desiccation. And Barth would surely have been even more surprised had he been able to observe the growing interest not only in continent-wide or even global climate changes but also in large projects to actively modify climates or—closer to the words of the engineers involved in these projects—to make the desert bloom again.[2]

I recount this history, tracing the upsurge of climatic anxieties among French and German colonial practitioners, before examining how some planners responded to these worries by proposing large engineering projects to counteract the perceived environmental decline. Europeans already had a long tradition of thinking about the effects of climates on both humans and their surroundings. From the late nineteenth century onward, however, they dramatically upscaled their climatic vision in two ways. First, practitioners now contemplated climatic changes affecting not just particular places and regions, but whole continents or even the entire planet; and second, engineers and planners used these ideas to design and justify large-scale, if ultimately unrealized, projects that would use the full force of industrial technology to halt or even reverse not only climatic but also societal and cultural decline. But what inspired Europeans to start thinking about extensive climatic shifts that could turn lush, forested environments into vast deserts? How, when, and where did engineers find inspiration in these theories to devise colonial engineering projects that attempted to transform those deserts back into forests? And how did the theories about both climate change and changing climates become embroiled in the political geography and the philosophies of cultural pessimism of the first half of the twentieth century?

In answering these questions, the book spans a wide arc, from the budding interest in large-scale climatic changes around the middle of the nineteenth century to Nazi plans to transform the climates and environments of war-occupied areas in the East. Connecting these chronologically and geographically distant places was the desert that Heinrich Barth had traversed in the 1850s. Whether conceptualized as a colonial repository for data, as a blank canvas for ambitious hydro-engineers, or as a symbol of environmental decline and desolation, the Sahara remained central to European, and particularly French and German, climatological concerns and designs throughout the period. First explorers and then climatologists theorized about its past and considered the possibility of widespread environmental and climatic changes that had shaped the history of landscapes and human habitation. The desert and its mutable ecology also inspired colonial officials and engineers to envision projects for an all-encompassing environmental and climatic transformation of the region or even the entire continent of Africa. Simultaneously, then, the Sahara was both a daunting representation of the awesome power of nature over humans and a surface on which to project designs that would reveal the newfound power of humans over nature in the era of industrial technology.[3]

Changing Climates

In the second half of the nineteenth century, popular science publications and magazines treated their readers to a continuous stream of reports by European travelers. These accounts told heroic stories about the adventurous feats of white men, but usually also provided some information on the inhabitants and environmental conditions of faraway places. Reports about Sahara travels habitually closed with references to a once lush landscape that had turned into a barren wasteland through desertification processes that were possibly still at work. The theory and apprehensions that formerly fertile or habitable land could have transformed into desert was not only a standard ingredient of European exploration stories but closely linked to the scholarly study of climates. As an independent academic discipline, climatology was still a field in the making but had already reached a sizeable audience and achieved a sturdy academic infrastructure. With a motley crew of practitioners from various related fields, climatology also featured a diverse set of approaches and methodologies, borrowing from geology, physics, and chemistry. More often than not, however, nineteenth-century climatic researchers relied on historic and geographic data to make inferences about climatic phenomena. While the methods they used were clearly different from the model-driven and computer-based climate science of today, one of the most important objects of investigation—climatic variation in the past, present, and future—was the same.

Western ideas about the variability of climates in the nineteenth century grew out of the confluence of new theories about the earth's geological past, imperialist expansion around the globe, desert exploration in North Africa, and the rise of geography and geology as established academic disciplines. Together, these developments prepared the ground for a lively scholarly discussion about the existence and possible causes of large, sometimes even global, climatic shifts in the past and present. The theories were as diverse as the evidence used to support them: from progressive warming to cooling, and from stable climates to short-term climatic oscillations. Eduard Brückner, one of the leading climatologists of the late nineteenth century, likened an overview of the competing theories to "walking through a veritable labyrinth without the benefit of Ariadne's thread." In the early twentieth century, the labyrinth was still just as difficult to navigate: the field remained without the stable framework of a generally accepted causal explanation that could have convinced a majority of the practitioners to adopt one among the many competing hypotheses. This did not mean, however, that large-scale climate change languished in the back pages of barely read journals. The debate elicited ideas and terminology that had already left the confines of academia and started to inspire public concerns over the possibility of environmental decline or even future climatic catastrophes that could destabilize colonial and even metropolitan economies and polities. Magazines published hyperbolic warnings of impending climatic doom, science fiction writers used climate change as both a setting and a plot device, philosophers ruminated on the connections between environmental and civilizational decline, and planners sketched afforestation schemes to counteract a looming environmental catastrophe.[4]

The emergence of ambitious engineering plans to counteract desiccation and desertification, whether regarded as man-made or "naturally" occurring, was one of the most striking manifestations of these growing climate anxieties. From the middle of the nineteenth century, some explorers, engineers, and colonial officials in arid parts of the world developed projects that would use the might of industrial technology to transform desert environments into fertile landscapes with climates fit for European settlement and agriculture. In the following chapters, I examine the designs proposed by three individuals and the implications of their projects: the French colonial engineer François Roudaire, who developed a plan in the 1870s to create large water surfaces in the Sahara to change climates; the pan-Europeanist architect Herman Sörgel, who proposed to completely reconfigure the Mediterranean to expand European climates, settlements, and culture to North Africa in the 1920s; and the German landscape architect Heinrich Wiepking-Jürgensmann, who pushed for his ideas of counteracting desertification through large-scale planning to be included in the official plans to Germanize the Nazi-occupied eastern territories during the Second World War.

Reflecting and refracting the debates on extensive climate change, the climate-engineering designs exhibited a deep environmental and cultural pessimism paired with a similarly powerful technological optimism, mixing narratives of crisis with those of redemption. Deserts—both as a powerful threat to be contained and as an inviting playground for the modern engineer's ambition—became coveted environments for climate modification projects. As different as the projects and their contexts were, their creators used and developed a common vocabulary forged in the climate change debate of the late nineteenth century, focusing on the Sahara as the quintessential desert in the European imagination. They also shared a deep techno-colonial impetus, promoting the use of modern industrial machines and tools to create productive landscapes for the benefit and habitation of non-indigenous populations, be they French, British, pan-European, or Germanic. To varying degrees, the schemes to convert desert environments became projects of social engineering. For the planners and their supporters, anxiety over encroaching deserts was always laden with cultural significance: environmental and climatic decline came to signify societal decline and vice versa. This added even more urgency to the fight against the purportedly encroaching deserts, but it also meant that engineering an amelioration of nature could potentially lead to an enhancement of culture or civilization in return—or at least to what the planners considered an enhancement.[5]

Colonial Science, Global Science, Desert Science

Both the climate engineering projects and the climate theories behind them were deeply embedded in colonial structures: explorers were on the payrolls of colonial offices, scientists used colonial infrastructure to conduct their research, and the engineers and planners of large-scale projects either worked directly for colonial governments or sought their favor and support. As Europeans occupied other lands during the age of High Imperialism in the last third of the nineteenth century, colonial officials paid greater attention to climatic conditions and, especially, to potential climatic changes in non-European environments. The possibility of climatic instability added to the sense of "unreadability" of little-known and environmentally alien overseas landscapes. Colonial planners also claimed climatic instability as a means to legitimize their colonial occupation and control, which involved deploying not only colonial agents but also colonial and colonizing science and technology.[6]

In many ways, this dynamic continued earlier colonial discourses and efforts to make colonial climates more agreeable and healthier for white settlers and administrators. At the same time, however, the nineteenth-century climate projects and theories were also part of the emergence of global science, a developing trend especially in geographical and geological disciplines, which built upon the collection and exchange of data from around the colonial world of the nineteenth and early twentieth centuries. In fact, the colonial and the global went hand in hand. Over the second half of the twentieth century, the data gathered by colonial governments and agents over decades gave modern climate scientists material to develop intricate numerical models of worldwide weather and climate forecasting. Today, climate science has become the most global science—or at least the most visible branch of global science—dealing with arguably the most far-reaching issue of all: anthropogenic global warming. But long before the last third of the twentieth century, climatology and climate anxieties had already scaled up to the global, with some European practitioners pondering and fearing the existence of environmental processes that affected the entire earth.[7]

Hypotheses about extensive climatic variability were part of public discourse long before the late nineteenth century—from Comte de Buffon's musings on climate amelioration in North America to Alexander von Humboldt's ideas about large-scale desiccation in South America. And yet scientists in the early nineteenth century still largely conceptualized climate as an essentially stable environmental feature that human action could only alter to a modest extent. This, however, was not to last. The gradual acceptance of the ice age theory around the middle of the nineteenth century opened the way to a reconceptualization of climate as a powerful and dynamic force shaped by and actively shaping environments from the poles to the equator. If, as proponents of the theory argued, the world had experienced substantial or even full glaciations in the past, what spoke against large climatic shifts in the more recent past, the present, and the future of the earth? Some climatologists, among them those working on extra-European, colonial environments, started to argue that the local or regional climatic changes they had described could be understood as part of much larger processes that spanned the entire globe. The exchange of climate data that took off in international conferences and journals in the second half of the nineteenth century aided the development of these large-scale theories of climate.[8]

Based on the musings about large-scale climatic phenomena, climatologists began to develop something resembling a global view of the environment, or what Mary Louise Pratt has called a "planetary consciousness." Theories about climate change—both anthropogenic and natural—played a central role in this development, as scientific practitioners increasingly portrayed environments as connected, inherently instable, and potentially malleable. Not just the Darwinian theory of evolution but also ideas and hypotheses of climatic variability and change made the nineteenth-century world and its environments appear increasingly variable or even unstable. In a recent study, Deborah Coen has traced the development toward a dynamic and multiscalar climate science in Central Europe over the last decades of the long nineteenth century, in which the varied environments of the Habsburg Empire provided the sites for research, data collection, and methodological innovation.[9]

The provinces of Austria-Hungary were certainly not the only colonial sites of climatological development. Some European practitioners interested in issues of climatic variability looked beyond their own continent and chose deserts as their primary field of research. With cave paintings, dry riverbeds, abandoned cities, and exposed geological features, arid landscapes provided an abundant source of evidence for long-term climatic changes. As powerful representations of danger, adventure, and desolation, deserts also played an important role in disseminating and popularizing ideas about environmental change and catastrophe. In the late nineteenth century, many of the deserts that climatologists studied, from Central Asia to Africa and South America, were either colonial or barely postcolonial environments and frequently represented territories that central governments situated in less arid places had claimed but not fully controlled.[10]

The colonial encounter with deserts was never a one-way street: as Europeans colonized desert environments, deserts also began to colonize European thought. This is visible not only in European ideas about and fears of deserts expanding but also in the work of early climate engineers, who often looked to create new Gardens of Eden in the arid zones of the world. They did not attempt to directly engineer the atmosphere, as some of today's geoengineers envision. Some early climate engineers nevertheless aimed to transform environments and climates beyond the local scale, designing lesser-known and unrealized, but certainly noteworthy, projects in colonial regions to halt the feared expansion of desert climates. These technological attempts to change environments and climates are a prime example of the early entanglement of climate science and colonial politics, representing the techno-political dimension of climate change ideas. While European practitioners often conceptualized deserts as empty spaces, climatological theories were never separate from the colonial contexts of their production. Rather, climate scientists took part in political debates and contributed to the search for solutions to perceived climatic and environmental issues in colonized spaces. Whether as proponents or critics of climate engineering, they frequently considered the possibility of continent-wide or even global climatic effects of human interventions. And climate scientists—along with climate engineers—theorized the effects of both natural and man-made climate changes on the social, economic, and cultural trajectories of both colonized and colonizing societies.[11]

Correcting Nature and Society

Climate engineering was not a new idea in the nineteenth century, and to many contemporary observers the concept did not seem out of the ordinary. After all, ideas about climatic instability were widespread among colonial planners and scientists. And if climates were indeed inherently unstable, the mental leap to attempting to change them intentionally was not that great—particularly with the tools of industrial technology and colonized deserts as testing grounds at the disposal of colonial engineers. Planners who proposed ambitious climate engineering projects in the nineteenth century were thus responding to mounting anxieties over environmental and climatic decline and expressing a growing belief in the possibilities of modern technology to reorder environments as well as societies. The scope of technological possibilities and the related scale of engineering projects reached new heights and new spheres of activity in the late nineteenth century. The German geographer Emil Deckert expressed the self-assurance of the age when he commented on a French climate-engineering project in the Sahara: "The unbound action of humankind [will] correct a number of critical mistakes of nature." And, as he added rhetorically, "to whom should this thought not appear beautiful and—if indeed feasible—tempting?"[12]

Deckert's words from 1884 seem to prefigure what is generally believed to be a recent phenomenon: the close link between climate change and macro-technology, which is exemplified by current geoengineering projects to halt

and reverse anthropogenic global warming. As Deckert's comment reveals—
and as I show in the following chapters—climate and macro-technology have
actually had a long-lasting relationship, reaching back deep into the nineteenth
century. In fact, climate engineering had already seemed tempting in the eighteenth century, when the Comte de Buffon declared that mankind would be
able to "alter the influence of its own climate, thus setting the temperature that
suits it best." In the nineteenth century this dream seemed to move into the
realm of feasibility: steam-powered machines and both metropolitan and colonial labor reserves provided new potential energy for large projects, while
the planetary perspective of environmental interconnectedness provided a
new scale for Western engineering ambitions.[13]

Deckert's comment is also noteworthy for its description of environmental
transformation. Rather than positing a contentious relationship between engineering and nature, Deckert, along with many of his contemporaries, emphasized the use of technology to "correct," or readjust, nature.[14] Once again,
engineering as a means to restore natural perfection and harmony was not a
new idea: earlier colonial climate improvement projects had often followed a
similar logic, attempting to restore local climates to an allegedly prior and
more perfect state. It is remarkable, however, that planners and commentators
in the late nineteenth and early twentieth centuries tended to frame even revolutionary projects aiming to create new climates, change geographies, and connect continents as conventional and, for lack of a better term, "natural." Engineers planning far-reaching interventions into environments and climates
tended to express their own role as that of a repairman tackling natural flaws
resulting from geological, cosmic, or anthropogenic processes. In the plans of
the early "geoengineers," nature and technology did not stand in opposition;
nor did they even belong to different conceptual frameworks. Climate engineers from the late nineteenth century to the first half of the twentieth often
saw the large-scale use of technology merely as a tool to perform a kind of
maintenance of nature, mirroring the effects of physical forces such as tides,
streams, winds, erosion, and climatic changes.[15]

While proposing some of the most extreme uses of industrial technology
in their time, early climate engineers also tended to not distinguish between
the "natural" and the "social" spheres, a separation that has had a firm grip on
some of the most durable conceptions of modernity and the technical imagination in the twentieth century. The separation was never complete and has
been questioned by both sociologists and historians. Nineteenth-century
ideas about climate change and climate engineering represent another powerful empirical argument against the overdrawn contrast between a stable "natural" and a dynamic "social" sphere. In fact, in the following chapters I suggest
that nature was at times brought more and more into the social sphere *through*

technology: after all, climate engineering proposed to offer mechanisms to steer and control climates similar to economic, political, and cultural parameters. François Roudaire, the engineer behind the Sahara Sea project, represents a striking example of the desire to revitalize and control nature as part of the French colonial project in North Africa.[16]

Conversely, climate engineers also voiced their hope that changing climates would engender social and cultural transformations. This was particularly visible in the first half of the twentieth century, when neo-Malthusian anxieties about the intrinsic limits to food production and widespread notions of civilizational decline combined with the fears of climatic change and desertification. Herman Sörgel's pan-European Atlantropa project, which aimed to transform European civilization alongside the climate of North Africa, was maybe the most grandiose embodiment of this dynamic, while the megalomaniac project of Nazi planners to comprehensively transform the environmental and racial characteristics of landscapes in the East was its infamous apogee.

The engineering projects that I examine, from the Sahara Sea in North Africa to the German *Generalplan Ost*, were more ambitious than any of their predecessors and among the most ambitious of their respective times. Their ambitions and their scale ultimately proved too large: the projects ended up unrealized, languishing in drawers in colonial departments or discussed to death in government offices and scientific journals. Rather than looking at the actual consequences of engineering projects as material, technological systems, I focus on the intellectual roots, the intended effects, and the impact of envisioned climate modification measures. While taking this approach forces me to stay in the realm of ideas and does not allow me to look at local consequences or responses, it provides insight into the most extreme forms of the technological imagination that were inspired by colonial and environmental anxieties from the late nineteenth century to the middle of the twentieth. If, as William Cronon has stated, "the nature inside our heads is as important to understand as the nature that surrounds us," then colonial imaginaries of environmental and social transformation are also as important as the everyday colonial encounters with non-European environments and their inhabitants. The proposed projects usually presupposed empty lands free of any indigenous habitation or activity and thus reinforced the colonial narratives of a tabula rasa, ready for European production and settlement. More generally, as David Edgerton has more recently pointed out, most innovations "fail" and are never used, but that does not mean that they are any less significant. The memory of their aspirations and their failures continues to influence new technologies and historical developments. As culturally embedded utopias or dystopias, unrealized designs—especially those seeking to transform vast swaths of the earth—thus play an important role in demarcating the extreme edges of the technological imagination. And that certainly includes large-scale climate engineering.[17]

Chapters

In the first chapter I explore how the discovery of ice ages, and thus the growing realization among European climatologists of the instability of paleoclimates, prepared the ground for a discussion of large-scale climatic shifts. In the second half of the nineteenth century, colonial travelers further fueled this discussion with information about environments and climates that they collected on their journeys in North Africa. Environmental knowledge about the history of the Sahara could serve both as data for scientific theories and as important information for colonial governments that wished to implement imperial projects. This was especially the case in the developing debate about climate fluctuations and climate changes that—as some European practitioners began to argue—had shifted the borders of the desert in the past and were still at work. The geographers and geologists involved in the debate presented a wide variety of theories and standpoints, but they moved largely between two poles: while one group held local, man-made causes responsible for climatic changes, others, and in particular German-speaking geologists and geographers, proposed that "natural" processes had been influencing climatic conditions all around the globe. By the early twentieth century the debate remained unresolved. This did not mean, however, that the topics of climate change and desertification vanished entirely. They reappeared periodically in scientific journals throughout the first half of the twentieth century. And the discussions about expanding deserts had left traces elsewhere: through the popularization of geographic and climatological knowledge, the issue of large or even global climatic changes and catastrophes had impressed itself upon the public imagination, while resourceful engineers had already started looking for ways to actively change climates through technological interventions.

In the second and third chapters I return to the nineteenth-century Sahara, exploring the emergence of colonial climate engineering projects. The second chapter examines a project by the French engineer François Roudaire, who developed a plan to flood a large portion of the Algerian and Tunisian desert in order to facilitate French access to the hinterland and, more importantly, to acquire new land for European settlement. The scheme was based on the premise that a large body of water in the North African inland would act as an evaporation surface, producing more precipitation and thus progressively altering the climate in the region. Even before the academic debate about climate variability had hit its stride at the end of the century, Roudaire took the idea one step further: he explored ways to engineer the climate and to return environmental conditions to their assumed prior conditions.

The third chapter looks more closely at the wider context of colonial climate engineering, exploring the reasons behind the sustained public debate about Roudaire's project. While commentators raised some doubts about particular features of large-scale climate projects, Roudaire's claim that he could bring about a considerable man-made change in climate often went largely uncontested. Like other similar projects of the time, the Sahara Sea project benefitted from a widespread Western belief in modern technology's ability to overcome all potential environmental and technological obstacles. It also tapped into both growing concerns about environmental decline among colonial planners and general anxieties about the precariousness of the French colonial project in North Africa. This mix of technological optimism and civilizational pessimism would become a common feature of climate engineering projects in the first half of the twentieth century.

The fourth chapter examines the development of this cornucopian-declensionist dynamic through one of the boldest successor projects to Roudaire's Sahara Sea, designed by the German architect Herman Sörgel in the 1920s. Atlantropa represented a gigantic plan to dam the Mediterranean and geoengineer a new combined Afro-European continent. Although Sörgel was working on a vastly different scale than Roudaire, his ideas were strikingly similar—and he in fact openly referred to his French predecessor's project as an inspiration for his own work. Sörgel's final goal for Atlantropa was to fertilize and colonize the Sahara by channeling enormous amounts of water through its midst and changing the climate to suit the needs of European settlers, while forcibly displacing the African population to the south. Sörgel was convinced that Atlantropa would bring about a new, peaceful era of progress and cooperation for Europeans and would form the basis for a coming postnational European society with room to expand on an engineered, colonized, and ethnically cleansed African continent.

In the fifth chapter I explore the transformative ambition of Sörgel's project, which took the idea of changing climates and geologies to an unprecedented level. Sörgel, who read widely not only in climatology and doomsday philosophy but also in geographical theories of the time, developed a model for overcoming cultural and environmental decline by changing the material conditions of their foundations or what could be called "active geopolitics." While this twist on geopolitical theories harmonized with some of the ideas of the new fascist government in Germany, Sörgel ultimately fell out of favor with the Nazi leadership. Sörgel focused all of his energies on colonizable land in the South, whereas Nazi planners looked to the East, where they discovered

Europe's own "Sahara" in the eastern steppes, which they deemed to be expanding just like its African counterpart.

The notion that Europe and Asia, in addition to Africa, were becoming successively drier had long been a theme in the debates among climatologists, especially in Russia. In the 1930s, so-called *Versteppung*, or "steppification," became a focus of Nazi planners. In the sixth chapter I trace the intellectual origins of work by Heinrich Wiepking, who became the main proponent of the *Versteppung* argument in the Nazi bureaucracy. In articles and books, Wiepking elaborated on the notion that once-fertile lands in the East had been turned into an arid steppe through Slavic settlement and possibly through larger climatic processes. These ideas would become central in preparation documents for the *Generalplan Ost*, which is my focus in chapter 7. The *Generalplan* sought to completely reorganize the East, combining ethnic cleansing with a comprehensive environmental and climatic transformation. Neither Wiepking nor his colleagues were particularly well versed in geology or climatology. They did, however, manage to incorporate and further popularize climate change anxieties that stemmed from the nineteenth-century debate. The Nazi administration used powerful images of *Versteppung* and desiccation to justify their military conquest and occupation of the East. *Versteppung*—stripped of all academic pretense—became first and foremost a political term charged with strong racist and fascist overtones during the Third Reich.

The epilogue connects the end of the *Versteppung* debate after the Second World War with our current concerns about global warming and desertification. I follow the development of climatology over the twentieth century, showcasing the discipline's split into global, atmospheric approaches in climate science and local, telluric approaches in soil science and desertification research. I end the book with an outlook on the rise of modern geoengineering schemes. Although these projects reflect current and presumed future technological capabilities, they also echo, perhaps unwittingly, terms, concepts, and fears of climatic catastrophe that have been around since the nineteenth century.

1

A Science of Sand

THE SAHARA AS ARCHIVE AND WARNING

AT FIRST SIGHT, the Swiss Alps seem worlds apart from the arid expanses of the Sahara. While both landscapes may trigger our Romantic sensibilities, the sensory contrast between the freezing cold of snow and ice and the glistening heat of desert sand is as great today as it must have been for nineteenth-century travelers. The two environments nevertheless share a long and entangled history in the annals of climate science, having served both as important research sites for scientists grappling with climatic issues and as symbols of environmental change.

In the iconography of anthropogenic global warming today, glaciers (shrinking) and deserts (expanding) together represent the alarming consequences of changing climates. In the age of "heroic science" in the nineteenth century, long before glacial ice cores and aridity indexes became central to climatology, European explorers had already become interested in extreme environments. They were drawn to these contrasting environments not only by romantic impulses but also by the search for answers to mounting questions about nothing less than the history and the future of earth itself. Deserts and glaciers provided—in fact embodied—material evidence of environmental change on a large scale. They opened a window to the distant, and possibly to the not-so-distant, past of the earth. Glaciers and deserts had moved and changed size—this slowly emerged as the scientific consensus. But how much had they moved and changed? And were they still in motion? In the middle of the nineteenth century, these inquiries engendered the first stirrings of a transnational debate on the existence and potential causes of climatic variability and progressive climatic changes affecting regions, continents, and even the entire globe. The debate was as scientific as it was societal. After all, the possibility of large climatic changes in the present immediately sparked expansive follow-up questions for scientific practitioners as well as officials and planners: what kind of consequences would climatic changes have on flora and fauna, on human settlements, on colonial projects, on the ever more intricately linked economies of the world, and ultimately on civilization? And could humans do anything about it?[1]

Theories and hypotheses about desiccation and deteriorating climates already had a long pedigree at that point. The late nineteenth-century debate, however, newly emphasized large-scale and global processes, the contextualization of climate change into a deep history of environmental transformations, and potential natural—rather than anthropogenic—causes for climatic change. By the 1870s, discussions of changing climates had become a common feature of geographical and geological publications, and they soon also found a popular audience beyond the academy. Despite ups and downs, false dead ends, and changes of emphasis, the debate about what caused unstable climates persisted and remained strong in the early twentieth century, with increasingly apocalyptic undertones. One particularly sensationalist article from 1905 warned of the "demon of the desert" that could potentially wreak havoc on all continents. Not all climate scientists agreed with this fatalistic view. Far from it, in fact; but both adherents and detractors were vocal about their views, bringing forward widely divergent and often contradictory views on the existence, periodicity, timing, scope, and causes of changes in climatic conditions.[2]

In these debates about desiccation and climate change, those who warned of "the demon of the desert" presented it in many forms, but most often in a North African guise. After Europeans began to explore the Sahara around the turn of the nineteenth century, the desert featured large in the work of European geographers and climatologists, whether as a site of research, an archive for climate knowledge, or a warning of the consequences of desertification. The desert moved even further into European consciousness once colonial powers began to formally occupy North Africa in the 1830s and sped up their colonial activity in the last third of the nineteenth century. The extremely arid landscapes of the Sahara and its border regions held valuable information for climate studies—not only in the form of geological evidence and quantifiable meteorological data, but also in ancient Greek, Roman, and Arab descriptions and physical traces of past environmental conditions, such as the ruins of once-flourishing settlements and rock paintings in the desert. As Europeans expanded their colonial occupation of desert lands around the world, they also debated how they might economically exploit these regions. And this would prepare the ground for a number of colonial officials, planners, and engineers to look at the Sahara not only as a lens onto past environmental conditions but also as a projection surface for large schemes to counteract environmental and climatic decline. At the beginning of this development, however, stood not the sands of the desert but the ice of the Alps.

A Science of Ice

In his 1840 study on glaciology, the Swiss geologist Louis Agassiz, commonly regarded as the founder of the ice age theory, stated that the fluctuations of glaciers were directly connected to the oscillations in global temperature in the past. As mundane as Agassiz's statement may sound today, it was revolutionary then. His recognition that the global climate had not always been constant in the past followed some of the more momentous geological findings of the first half of the nineteenth century, first among them the gradual dismissal of an inherently stable and immutable earth.[3]

The first stirrings of this shift in the European worldview had already taken place in the late eighteenth century, with authors from David Hume to Pehr Kalm and Jean-Baptiste Dubos alluding to large climatic shifts. The famous historian Edward Gibbon also joined the fray by drawing a correlation (if not a direct causation) between the fall of the Roman Empire and changes in climatic conditions. Beyond the arguments of individual writers, a broader intellectual development would become even more important for new geological theories: carefully at first and then with ever bolder arguments, some scholars started to break with the traditional biblical chronologies that had limited the age of the earth to the meticulously calculated divine creation. Starting in the first decades of the nineteenth century, then, glaciers became important sites for the elaboration of ice age theories and for the discussion of enormous geological and climatic changes in the newly discovered deep past of the earth. Around the middle of the century, Louis Agassiz became the most famous proponent of these theories.[4]

Agassiz was a latecomer to the study of glaciers. In the early 1930s, he had still openly rejected some of the glaciation theories suggested by his colleagues. And even when he did start to work on his own ice age theory a few years later, it was decidedly idiosyncratic: Agassiz believed that large parts of the earth had been covered with glaciers *before* the Alps had formed and that giant ice sheets sliding off the rising mountains could explain the material evidence of glacial motion in Alpine valleys. The theory made sense against Agassiz's understanding of the history of life on earth. Faced with the evidence of fossil remains of extinct organisms, but firmly holding on to the idea of a divine origin of species, Agassiz could use the theory of glaciations to serve his own antievolutionary beliefs. Thus, he argued, the earth had been undergoing a general process of cooling, which had been happening not in a continuous downward slope but in discrete steps of periods marked by constant climatic conditions, separated by short bursts of rapid cooling and subsequent phases of partial warming to the next stable climatic plateau.[5]

This peculiar model solved two of Agassiz's intellectual quandaries at once. First, it harmonized with ideas of a gradually cooling earth, which had been proposed by the French mathematician Jean-Baptiste Joseph Fourier in the 1820s and had found a broad audience since then. Even Charles Lyell, the most famous geologist of his day, had to make room for the apparent temperature and climatic changes of the earth in his "doctrine of absolute uniformity," the hypothesis that the geological past had experienced the same kind and intensity of physical events and processes that occurred in the present. Second, Agassiz's model allowed him to explain how organisms enjoyed stable environmental conditions up to an eventual event of mass extinction caused by a catastrophic drop of temperature, which was followed by the divine creation of new species for the next period of stable temperatures. Agassiz now had a way to explain mass extinctions without turning to evolutionary theory.[6]

Seemingly buoyed by this expedient conjunction, Agassiz began to devote his energies to expounding the details of his ice age theory. In the late 1830s, he extended the proposed prior extent of the earth's glacial cover to encompass large parts of the globe from the North Pole to North Africa. In a letter to the English theologian and geologist William Buckland, Agassiz—declaring himself to be "of a quite snowy humour"—even referred to the "whole surface of the earth covered with ice." For his contemporaries, however, this was a lot to swallow: after all, Agassiz's take on the glacial theory not only upset uniformitarian notions of a steady-state earth with slow cyclical patterns of change but also did not sit well with catastrophists, who found it difficult to align their commitment to sudden, noncyclical changes with Agassiz's assertions of a general, very slow trend toward a cooler earth and the periodic nature of rapid cooling and warming events.[7]

Despite the slow uptake of the glacial theory, a seed had been sown. Over the next thirty years, evidence about more extensive early glaciations from various parts of the world continued to accumulate. Gradually, Agassiz's ideas about a protracted ice age, if not his more idiosyncratic ideas about multiple creation events, gained more widespread acceptance. Eventually, the glacial theory, positing both drastic and repeated changes in the earth's past, triggered first a crisis and then a revolution in geology. The postulated occurrence of ice ages in the *recent* geological past, in particular, forced geologists to grapple with a new history of the earth marked by contingency and instability. Past changes in the geological and climatic conditions of earth now became genuine objects of scientific study. Large environmental and climatic changes in the recent geological past also meant that these same kinds of processes were possibly still at work in the present and could affect the future of the globe.

Of Cooling and Warming

While the ice age theory was still struggling for general acceptance in the middle of the nineteenth century, Agassiz's writings on large climatic shifts had given a new impetus to the debate about a cooling earth, which Fourier had postulated in the 1820s. One of the most influential participants in the discussion was the British physicist William Thomson, better known as Lord Kelvin. Inspired by Fourier's work, Kelvin began his own studies of the dissipation of heat in the 1840s. He undertook calculations to gauge the ages of the earth and the sun from the estimated amounts of dissipated heat of the two celestial bodies. Kelvin's results, assessing the age of the earth at under two hundred million years and that of the sun at possibly less than one hundred million years (Kelvin adjusted these numbers at various times), did not sit well with geologists, many of whom were already thinking in much more expansive, or even infinite, time scales.[8]

Kelvin's ideas, particularly his emphasis on the potential consequences of a general cooling trend for humanity, nevertheless made an impression. The sun, Kelvin wrote in 1862, was still powerful enough to maintain life on earth but would eventually expire: "As for the future, we may say, with equal certainty, that inhabitants of the earth cannot continue to enjoy the light and heat essential to their life for many million years longer unless sources now unknown to us are prepared in the great storehouse of creation." While the doomsday scenario that Kelvin painted here lay far in the future and Kelvin himself did not play any manifest role in the climate change debate that was to unfold, his ideas helped to spark heightened interest in large environmental transformations.[9]

Around the same time of Kelvin's publications, Agassiz's friend and colleague John Tyndall worked on his own development of some of Fourier's ideas, especially on the insulating qualities of the earth's atmosphere. Tyndall examined the absorption of radiant heat by different gases, which would earn him twentieth-century fame as one of the discoverers of the greenhouse effect. In the 1890s, his work would also inspire Svante Arrhenius's studies on the impact of atmospheric carbon dioxide on temperature and climate, which complicated Kelvin's theory of a steadily cooling earth. Arrhenius, who is now often placed at the origin of histories of global warming research, suggested that changes in the amount of carbon dioxide in the earth's atmosphere could have caused the Ice Age. Almost sixty years after Agassiz had postulated a glacial epoch, there was now a mechanism that could explain large climatic variations in the earth's past, although the theory found only few adherents and went largely unnoticed until the mid-twentieth century. Back in the late nineteenth century, Arrhenius was already thinking about how humans could impact the concentration of atmospheric carbon dioxide by burning fossil fuels. In contrast to current concerns, however, Arrhenius actually welcomed the warming that continued coal combustion would cause.[10]

Arrhenius's favorable view of global warming was not out of place in the nineteenth century, especially not among his neighbors in cold Scandinavia. As geologists started to accept the theory of one or more ice ages in the past, as well as hypotheses of a cooling earth, they began to imagine drastic climate changes in the present or future that could lead to a renewed glaciation of the earth. Agassiz's insistence in 1840 that there had been no appreciable climatic changes in historical times—a point congruent with his particular model of climatic patterns in the history of the earth—did not remain unquestioned over the course of the century. Popular scientific publications from Europe to North America warned of imminent glacial epochs, or at least of large cooling events in the near future.[11]

One 1887 article, printed in the French *Revue scientifique* and translated for the *Scientific American,* dealt with what is now known as the Little Ice Age and referred to substantial geological evidence that the world had entered a new glacial cycle in the thirteenth century. The agreement among scientists was certainly not as universal as the author suggested. Similar arguments nevertheless cropped up in various scientific and popular publications in the second half of the century. Some authors focused on water currents and wind patterns, which appeared to better explain more rapid changes in climate than the slow burning up of the sun or major variations in the atmosphere's composition. A popular 1889 book on meteorology suggested that a diversion of the Gulf Stream could easily initiate a new ice age in Europe. And just a few years later, another article pointed to ominous developments under way near the South Pole, warning that the growing icebergs and glaciers were a sign of a new glacial epoch.[12]

Robert MacFarlane has suggested that the return of an ice age was the Victorian version of a "nuclear winter." And with terms like "glacial nightmare" making the rounds, the apocalypse steadily accompanied discussions in the popular scientific press about the possibility of future glaciations. Tyndall, himself an avid mountaineer and glaciologist, wrote suggestively that the "glaciers of the present day" were "mere pigmies as compared to the giants of the glacial epoch." Depictions of glaciers—both verbal and visual—spread from the Alps through the growing transnational networks of science and the popular press, making it into journals, magazines, and newspapers across Europe and the United States. At the same time, the rise of mountaineering opened up a new playing field for the expression of rugged manliness and a special bond to nature, bringing even lowland Europeans into closer contact with the mountainous environments of the Alps and disseminating images of snow-capped peaks and glaciers to an ever wider audience.[13]

With their novel theories about the geological history of the earth, Agassiz and his colleagues had helped to create a more general understanding of the malleability of nature—a nature that now came to figure in the scientific imagination as both much older and much more unstable than previously imagined. Together, evolutionary theories and glaciology informed how nineteenth-century Westerners reimagined the earth as a dynamic and changing place. The glacial theory was thus important for providing new ways to conceptualize geological and climatic transformations in the deep past but also for opening up the possibility of large-scale environmental changes in the historical past, the present, and the future. The apprehensions expressed in the popular press over a return of the Ice Age due to the earth's cooling were only *one* product of this development. The possibility of the opposite process, namely the gradual warming of the globe or at least large parts of it, was another new source of research and anxiety.

Once again, Alpine landscapes could provide an inspiration for this inquiry into climatic change, as was the case for John Ruskin. The Victorian art critic and social thinker attempted to capture the awesome beauty of glaciers in watercolors and thought about the past and future of glacial landscapes (Figure 1.1). In fact, Ruskin debated heatedly with John Tyndall about the rate of glacial retreat, thought about the effects of deforestation on climate, pondered the impact of atmospheric pollution, and warned his audience about the societal and environmental dangers of a fossil-fuel-based economy. For Ruskin, in fact, all of these ideas and activities seemed interconnected: climatic decline reflected a more general societal malaise—a theme that runs like a red thread through the ideas about changing climates in the nineteenth and early twentieth centuries.[14]

Through the Sahara

Just like icy glaciers, deserts became important locales for stimulating both research and anxiety among scientists, explorers, and colonial officials. Even long before the nineteenth century, deserts had held a particular allure for Europeans. As places of refuge, religious purification, and exile, or as the monster-infested borders of the known world, they had been significant locales in the imagination of writers and readers. In the nineteenth century, deserts moved even further into the meeting rooms of scientific associations, into university lecture halls, and ultimately also into the living rooms of the bourgeoisie. The Sahara, as the desert closest to the European mainland, emerged as a central research site for the budding field of climatology around the middle of the nineteenth century. The new level of scientific interest in desert environments mirrored the new level of colonial interest in non-European

FIGURE 1.1. John Ruskin, *Glacier des Bois* (1843).

lands during the age of imperialism. In fact, scientific, economic, and political concerns were inextricably connected, which was nowhere more evident than in the expeditions sponsored by European governments and colonial associations. Heinrich Barth was one of these travelers, and a few of his observations would become some of the most important resources for practitioners interested in large-scale climatic changes in the second half of the nineteenth century.

From 1850 to 1855, Barth crisscrossed Africa, reaching the shores of Lake Chad and paying a visit to the holy grail of Sahara travel: the legendary city of Timbuktu. Unlike many of his predecessors and his own European travel companions, Barth survived to tell his story. Equipped with training in Arabic, a deep knowledge of Islamic theology, considerable diplomatic skills, and a substantial amount of good fortune, he survived various altercations and diseases

FIGURE 1.2. Artistic representation of Heinrich Barth, dressed
as an Ottoman doctor, among North and sub-Saharan
Africans. *Source*: Robert Brown, *The Story of Africa and Its
Explorers* (London: Cassell, 1892).

in the Sahara. All of this should have made him a star of nineteenth-century imperial culture, his name inscribed in the popular annals of exploration and "heroic science" next to giants like David Livingstone and Henri Duveyrier. Even before the end of the century, however, Barth's name had been all but forgotten by the German public and lived on only within the scientific community.[15]

Barth's failure to reach the airy heights of nineteenth-century heroism was partly due to the timing of his journeys, which began shortly after the 1848 revolutions and ended well ahead of German unification in 1871. Before German imperial interests could find or form an institutional base, Germanophone explorers often sought support from governments and organizations in other countries. Since the late eighteenth century, the London-based Association for Promoting the Discovery of the Interior Parts of Africa, generally called simply the African Association, had supported travelers in the Sahara region, among them the famous Scottish surgeon and explorer Mungo Park. It had also funded the fateful journey of German expedition leader Friedrich Hornemann, who disappeared in the northern Sahara in 1800. Fifty years later, the Royal Geographical Society in London absorbed the African Association and continued its work by publishing the interim reports sent from North Africa by Barth and his travel companions, the British missionary and abolitionist James Richardson and the German botanist Adolf Overweg. Once returned from Africa, Barth moved to London, where he set about writing the full report of his voyage, which first appeared in 1857 in an English edition.[16]

Barth's 1857 travelogue was the second, and maybe the more important, reason both for his failure to become a popular icon of the age of exploration and for his enduring appeal among the scientific community. Written in a dry style and published in five dense volumes, the report provided little entertainment but copious facts on the botany, zoology, geography, geology, and—perhaps most prominently—ethnography of the Sahara region. While this onslaught of information tended to overwhelm or bore the casual reader, it did inspire a whole generation of European explorers and scientists who would add to the Western body of knowledge about North Africa over the following decades. Two years after Barth's *Travels and Discoveries in North and Central Africa* was published, the aged Alexander von Humboldt had already realized its significance for the study of the Sahara region and praised his far younger colleague as "having unlocked a continent for us."[17]

Barth's work certainly helped to inspire scientific studies in the Sahara, including climatological investigations. Even on his preliminary trips through the coastal areas of the Mediterranean, Barth had reported on water sources that appeared to have once had a far higher yield. Standing among ruins of former habitations, he showed his disappointment at the marked decrease in fertility. "Once," Barth wrote about the area around the Libyan Gulf of Bomba, "a perennial water might have livened up everything here, spreading freshness and abundance, where now dearth and the burning sun enervate the traveler both mentally and physically." During Barth's first crossing of the Sahara, indigenous guides pointed him and his expedition companions to rock drawings in the Fezzan, the extremely arid southern part of modern Libya (Figure 1.3). To the explorers' surprise, the images depicted, among other scenes, cattle and antelopes—a strange sight in an environment that at the time lay clearly outside the habitat of any large mammals. Barth began to wonder about possible changes in the climate that might have desiccated the area, pushing the habitats of large wildlife further to the south. In his 1857 report, he referred to this process with a casual, but noteworthy, remark about a general increase in aridity throughout the entire Mediterranean region.[18]

FIGURE 1.3. Rock drawings in the Fezzan. *Source*: Heinrich Barth, *Reisen und Entdeckungen in Nord- und Central-Afrika in den Jahren 1849 bis 1855*, 5 vols. (Gotha: Justus Perthes, 1857), 1:210, 214. Barth had to reconstruct the images from memory, as the reproductions he drew on the spot got lost during his travels.

Barth's successors, who would add to the European knowledge of the Sahara, picked up the thread and continued the search for evidence of a once fertile Sahara, using historical, archaeological, geological, botanical, and sometimes even zoological evidence. By 1876, Paul Ascherson, the German botanist and travel companion of the celebrated Sahara explorer Gerhard Rohlfs, could already refer to a standard repertoire of facts about climatological decline in the desert, which, he argued, could not be attributed just to the oft-cited neglect of antique irrigation works under Arab rule. Other, more powerful processes had to be at work as well. A few years later, Ascherson and Rohlfs became embroiled in a debate about lions inhabiting the Sahara, which had been started by the German traveler Erwin von Bary. In a letter to Ascherson, Bary suggested that lions and other large mammals had still lived in the northern parts of North Africa in the recent past. After Rohlfs brashly critiqued the argument one year later, Ascherson came to Bary's defense by arguing that neither of them had ever argued for the *present-day* existence of lions in the Sahara, thus implying a large and relatively recent shift of environmental conditions that had pushed the habitat of lions to less arid regions south of the Sahara.[19]

While the dispute itself was short-lived, it exemplifies the interest of explorers in environmental and climatic variation in the Sahara. Rohlfs, who dismissed the idea of lions in the Sahara, still believed that large climatic changes had happened in Africa in historical times. And despite Rohlfs's critique, Bary's hypotheses remained influential. In his geological survey of the Sahara, the paleontologist Alfred von Zittel cited traces of large mammalian life, alongside rock paintings, as evidence for climatic changes in the region. Not local geological events, he concluded, but climatic changes had transformed the once water-rich northern half of Africa into desert. Reports from Sahara travelers, among them such familiar names as Henri Duveyrier, Gustav Nachtigal, and Georg Schweinfurth, remained centrally important in the debate over climate change, which began to pick up steam in the 1870s. Even Barth's travel accounts from the 1850s continued to be cited. The geographer Theobald Fischer, one of the most influential and most-cited participants in the debate, repeatedly highlighted the importance of Barth's writings for the debate on historical changes in the climate of the southern Mediterranean region. And in 1909, more than fifty years after Barth's trip had concluded, the authors of a review of climate change in North Africa still referred to Barth's reports as one of the study's most important sources.[20]

More often than not, the information on environmental changes brought across the Mediterranean Sea by Sahara travelers was not actually firsthand knowledge. However heroically the European explorers portrayed their own exploits, they were almost completely dependent on indigenous North Africans as "go-betweens," who acted as guides, translators, mediators, and armed protection, as well as scientific informants and practitioners. Richardson, Barth, and Overweg would not have "discovered" the rock paintings, nor have even reached the southern Libyan Desert, without local guides, whose histories usually went unrecorded in the publications and archives of European travel in the Sahara.[21]

The physical evidence in the Sahara—from drawings and ruins to geological formations—could point to climate changes in a rather unspecific past, reaching back hundreds or thousands of years in the case of traces of human habitation, and even millions of years in the case of geological evidence. Only the stories and accounts of the indigenous population could supply information on more recent changes. The European travelers in the Sahara, cognizant of their own limitations and dependence, often hid this behind a façade of superiority. Thus, in an account of a journey from Morocco to Tripoli, Rohlfs advocated for exterminating Arabs in North Africa and creating an all-European Algeria, all while he was staying as a guest in Arab homes and relying on North African guides and porters. And despite his appraisal of the resident population as generally useless and untrustworthy, Rohlfs was happy to receive and use local information on the frequency of rains in the recent past.[22]

Like their counterparts in other disciplines grappling with a shortage of reliable data, climate scientists in Europe knew the limits of available information on climate change but still only begrudgingly used eyewitness accounts by non-Europeans. With the age of "heroic science" drawing to a close, this stance became even more categorical in the early twentieth century. In 1909, the Austrian climate scientist Hermann Leiter doubted indigenous knowledge outright, claiming that North Africans did not have any clear notion of how the climate had changed and would simply say whatever their European interlocutor wanted to hear. Another publication in the same year deemed as unreliable the accounts of the colonized, as well as those of missionaries.[23]

Europeans' mounting dismissal of local informants had two distinct points of origin. First, by invading North Africa the colonial occupiers not only had easier, direct access to research sites in desert environments but also sharpened their perceptions of racial and cultural difference between "civilized" Europeans and "uncultured" Africans, which contributed to the devaluation of local knowledge. Second, climate scientists could not overcome the persistent dearth of quantifiable data on the North African climate, which led some to become frustrated, blaming and criticizing local informants for their inability to unearth scientific information. The lack of comparable data series grew to be all the more embarrassing toward the end of the nineteenth century, when some practitioners in the young field of climatology were starting to abandon their reliance on eyewitness accounts and attempted to stake out a claim for climatology as an independent and exact science. This process also signaled the beginning of a move from a telluric, or earth-bound, understanding of climate, based on observation and description, toward a physics-based conception of climate and a new focus on atmospheric phenomena in the early twentieth century.[24]

But it was a long way to this point, and the climate change debate in the late nineteenth century was first and foremost a geographical endeavor. While environmental and climatic changes, and particularly desiccation, had long been discussed in colonial circles, the institutionalization and the concurrent popularization of geography made it possible for climate change theories to reach a wide and receptive audience. Closely tied to colonial exploration and expansion, the rise of geography as both an independent academic discipline and a bourgeois pastime provided an institutional home and an avid readership for climatological debates. In the volatile period of the reclassification and institutionalization of knowledge in the nineteenth century, climate studies emerged first as a shared subfield of geology and geography, and then, toward the end of the century, more and more as an independent discipline with its own journals and conferences, vocabulary, and standards of data collection and interpretation. Throughout this development, the question of the existence of large-scale climatic shifts remained both one of the intellectual driving forces and a constant bone of contention.[25]

Climates Changing

"Knowledge is power—geographical knowledge is world power." This was the candid motto of the Perthes Verlag, a German publishing house that would become the most conspicuous mouthpiece of the geographical community in German-speaking lands. The motto tied the pursuit of geography inextricably to colonization. And, indeed, over the second half of the nineteenth century, geography became the foremost auxiliary science of European colonialism, with geographers equipping armies and bureaucracies with maps of overseas territories and information on environmental conditions. The nineteenth-century climate change debate became a regular feature in geographical journals, and modern climatology emerged as imperial climatology in this context. Data on climate became an important point of reference for imperial governments evaluating economic risks and potentials in overseas colonies. Particularly in arid regions, the amount and, even more importantly, annual pattern of rainfall and the extremes and averages of temperatures were vital for the assessment of agricultural prospects and projects to acclimatize nonnative livestock and crops.[26]

Despite—or maybe because of—this new importance of climatic studies, its place within the emerging disciplinary landscape of the late nineteenth century remained unresolved. While geologists had been dealing with questions about large climatic shifts since the ice age debates and colonial geographers laid claim to include climate in their holistic descriptions of overseas environments, some practitioners started to call for the scientific community to move away from describing landscapes and their characteristics qualitatively and shift to studying climates more quantitatively. This development was made possible only by the fast-expanding network of meteorological stations that had been recording data series on weather conditions—starting first in Europe and then, without much delay, in the colonies. By the 1870s, most European nations had set up state-sponsored meteorological services that collected quantitative data series on weather conditions. And by the end of the decade, meteorology and, to a lesser degree, climatology had become institutionalized themselves, as evidenced by textbooks, journals, international conferences, and, most conspicuously, the foundation of the International Meteorological Organization (now the World Meteorological Organization) at the First International Meteorological Congress in Vienna in 1873.[27]

Despite the rapidly developing meteorological infrastructure, climate data beyond the immediate past and beyond the borders of Europe remained scarce. Alongside the physical evidence of ancient ruins and rock drawings, historical texts were still the most popular sources of information on past climatic conditions in North Africa, especially Greek, Roman, and Arab accounts. The historical depiction of Mediterranean environments, often depicted as more fertile and wetter than present conditions, had been a critical inspiration for research into climatic conditions in the first place. According to some accounts, Barth took Herodotus's travelogues as his only reference book on his journey into the Sahara. Herodotus's *Histories*, which quickly became the most cited source among participants in the North African climate debate, portrayed a settled and fertile North Africa. He described antelopes and other large animals, wrote about large lakes, and told of thriving cities, although he also portrayed the noncoastal areas of North Africa as a big barren desert devoid of any life. The question for climate researchers, however, had never been whether the Sahara had existed at all in the past but whether its borders had shifted in historical times. And Herodotus's accounts of a thriving coastal North Africa seemed to give some indication that they had.[28]

The evidence was even more convincing in combination with the accounts of other classical authors, such as Pliny, Procopius, Strabo, and Ptolemy, which corroborated a once more fertile North Africa. Some climate scientists also turned to the accounts of the Arab historian Ibn Khaldun, who described North Africa's environment in the fourteenth century and had already been an important source for eighteenth-century Enlightenment ideas about climate. And then there was also the common notion of North Africa as the "granary of Rome" that had supplied the populous ancient empire with grain over centuries. The arid lands that European travelers, soldiers, and colonial officials encountered in North Africa seemed a far cry from the fertile lands of the past. What had happened? Starting in the 1870s, this became the central question for climate scientists, who advanced a number of conflicting ideas and hypotheses.[29]

The idea that the North African environment had undergone drastic environmental and climatic changes never garnered unequivocal support and remained contentious at the end of the nineteenth century.[30] Even among the adherents of a climate change hypothesis, opinions differed widely on the extent and cause of the changes. While some argued for local changes, others now advanced ideas of global climate shifts or variations. And while some argued that changes were man-made, others started to see geological or even cosmic processes at work. Beyond that, practitioners also disagreed on whether the climate changes were progressive, namely tending toward more extreme temperatures and conditions, or cyclical, displaying recurring ups and downs of varying lengths. All in all, the climate change debate of the late nineteenth century was wide-ranging and sometimes simply confusing. Looking back at the previous half century of the debate in 1909, Wilhelm Eckardt wrote in his aptly titled study on the "climate problem" that the literature on climate changes in both the geological and the historical past was "extraordinarily rich, but also very scattered." Eckardt's statement demonstrates that even in the early twentieth century, the climate change debate had not arrived at anything resembling a consensus. Despite the cacophony of voices and opinions, however, some patterns in the debate allow us to reconstruct the central strands of thought and main dynamics.[31]

The onset and the early development of the climate change debate coincided almost perfectly with a new focus on declensionist narratives among French colonial administrators in North Africa. While they had been worried about the state of the environment as soon as French troops occupied Algiers in 1830, a fully developed account of environmental decline and desiccation showed up in reports and articles only from the 1850s onward and then became widespread in the 1870s. French colonial officials adopted this narrative as a rationale for colonial expropriation and stewardship of the land, in order to support their continued and expanding presence in North Africa. They also used the narrative to justify strict measures of economic and social control over the indigenous population, whom they blamed for the alleged environmental damage through the destruction or neglect of irrigation works and widespread deforestation.[32]

This view was endorsed by some climate scientists and many other commentators in different fields of expertise, both in France and beyond. In

Germany, the theory of an ongoing desiccation of North Africa found an early adherent in the agronomist and botanist Carl Nikolaus Fraas, whose 1847 study used plant geography and the traces of past agricultural production as evidence to substantiate claims of large and widespread climatic changes—and a continuing desiccation—in Egypt during historical times. His namesake, the geologist and clergyman Oscar Fraas, supported that view and connected the cultural decline of Egypt to the higher average temperatures and lower precipitation in contemporary North Africa. During the apex of Egyptian civilization, he argued, the desert simply could not have had its current extent, as high cultural development required wetter conditions than the current climate provided.[33]

Oscar Fraas's view of the derelict state of both the North African environment and its culture corresponded to French colonial discourses of environmental and civilizational decline. While many French, British, and American writers leaned toward an explanation of climate changes through human action such as neglect, deforestation, or overgrazing, however, Fraas gestured toward a different explanation. He admitted that much of North African history would probably remain unexplained, but ascribed the desiccation of the region to "geological oscillations of elevation and cosmic transformations." Fraas remained vague in his description of the mechanism behind these processes, but his proto-explanatory model fit the general thrust of the debate among Germanophone climate scientists. Overall, writers trained at German-language universities tended to foreground, or at least discuss, "natural" causes of large climatic changes in historic times. Rather than focusing on the destruction of forests and soil, they often referred to ill-defined geological, solar, or cosmic processes to explain apparent environmental changes in North Africa.[34]

The origins of this predisposition are not entirely clear. To be sure, glaciological research into large-scale climate changes in the geological past continued in the second half of the nineteenth century and may have influenced ideas about more recent climatic changes. This, however, would not explain why this preference had a decidedly *German* dimension, as both the primary research site of the Alps and the background of glaciologists stretched beyond national and linguistic barriers. At least in part, the preference in Germany for non-anthropogenic explanations may therefore have been connected to the nation's lack of direct political influence in North Africa and the Mediterranean region—a state of affairs that not only continued after the Berlin Conference of 1884 but was in fact cemented by the ensuing division of the continent into European spheres of influence. At least until the halting beginnings of a German settler colony in Southwest Africa in the last decade of the nineteenth century, German-speaking scientists had less of a vested interest in legitimizing narratives of indigenous neglect and destruction in arid overseas environments. This did not mean that German scientists turned categorically against anthropogenic explanations of environmental change or that it made them more favorably inclined toward indigenous populations. It did mean, however, that in their climatic analyses Fraas and his colleagues often looked beyond the centers of colonial power in the relatively densely populated coastal areas of North Africa to include the more extreme and less economically promising environments of the desert interior.[35]

Theobald Fischer exemplified this "German" approach. Originally trained as a historian, he soon turned to geography in an attempt to connect his humanistic training to the natural sciences. His work focused on the Mediterranean region, which allowed him to combine methods and approaches from classical exegesis with meteorological analysis. In his writings on North Africa, in particular, Fischer emphasized the large and widespread environmental changes that had happened in the recent past. While he supported the view that the coastal areas of North Africa had suffered a harmful deforestation and subsequent desiccation, he diagnosed a "general telluric process" of desertification and climate change south of the thirty-fourth parallel. "In many regions," he wrote, "it is possible to demonstrate directly . . . that a considerable decrease of precipitation has taken place since ancient times, and is apparently still taking place in the present at such a high rate that vast areas have become uninhabitable for a sedentary population." Citing the ruins of settlements in the western Sahara that were now located in the midst of the desert, Fischer went directly against anthropogenic explanations of climate change and contended that the desertification of the region could not solely be ascribed to historical processes but had to be attributed to meteorological processes. This was the case not just in Africa, Fischer argued, but everywhere around the globe along the same latitude.[36]

In these statements, Fischer touched on three major issues that would come to define the climate change debate. First, he characterized the Sahara as a relatively young desert and attested to a continuing and ineluctable process of "desert formation"—a bold viewpoint that made him known and widely cited within and beyond German language borders. Second, Fischer mentioned the human ability to counteract desiccation and climate change, if only to say that this capacity was restricted to particular places and circumstances. And third, Fischer gestured toward large-scale climatic changes, linking processes of environmental change on different continents and contributing to the emergent conception of a "global," or at least transregional and transcontinental, environment.[37]

While Fischer took a rather cautious stance on the human capacity to modify climates, some of his contemporaries were already far more optimistic and entertained visions of transformed desert environments, often citing the benefits of planting trees. Whether afforestation could combat changing climatic conditions became one of the most controversial topics in the debate. While colonial governments were already busy growing eucalyptus forests in arid regions, not everybody was convinced that these measures would ultimately succeed in pushing back deserts. The advocates of tree-planting programs often subscribed to the causal connection between the destruction of forests and climate change. The almost unavoidable reverse logic was that afforestation would be the best countermeasure. Others, including Theobald Fischer himself, were more careful in their evaluation. While forests could moderate climatic conditions, they argued, not all environments were sufficiently water-rich to support forest growth. Eckardt, in 1909, summarized the inconclusive results on the effect of forests, which on the one hand appeared to bring about a "moderation of temperature extremes," but on the other hand seemed to have only a negligible effect on air temperature. Overall, he concluded, forests could have a local effect on climatic conditions but were not able to influence the temperature of larger regions.[38]

In the nineteenth century climate scientists made larger regions—sometimes whole continents or even the entire planet—a common object of inquiry. This was the third major issue that Fischer touched upon in his writings. While he restricted his own research to the Mediterranean region, Fischer referred to desiccation in desert zones all around the earth. Trained in the synthetic discipline of geography, nineteenth-century climate scientists often attempted to translate their knowledge of regional climates to a more holistic understanding of transregional or even global climatic trends. This was evident in some studies that attempted to compare climatic conditions in different regions and found similarities and common patterns. Thus, an Austrian study on the "variability of the climate" found parallel trends of desiccation in arid regions from Palestine to the Hungarian steppes and a tendency toward cooler temperatures all around the North Polar Circle, although—following the general fashion among climatological work of the period—it shied away from volunteering an explanation for these phenomena.[39]

The German climatologist Eduard Brückner was perhaps the most outstanding example of this "global" trend among climate scientists. He believed—mistakenly, as it turned out—that his "discovery" of worldwide climate oscillations in the 1880s would end the debate on progressive climate changes. In his attempt to show the universal applicability of his thirty-five-year climate cycle model, Brückner collected eclectic climate data from all around the world. That this global repository of climate data was even possible, if still challenging, was due to the fact that the climate change debate had quickly spread beyond North Africa. This was in part a consequence of the frequent, if often unsubstantiated, intimations of global climatic changes, which inspired data collections and climatological work in different arid regions around the world, in South Africa, in North America, and—with particular vigor—in Central Asia.[40]

In 1904, an article in the *Geographical Journal* by the Russian Pjotr Kropotkin—better known as an anarchist than a geologist—triggered an international discussion on the "Desiccation of Eur-Asia." With evidence of falling water tables and shrinking lakes in Central Asia, Kropotkin argued that Asia was undergoing rapid desiccation, which could not be explained by the anthropogenic destruction of forests alone. Larger forces were at work, he wrote, and these forces had been active since the end of the last ice age and now threatened the entire northern hemisphere with an unrelenting advance of desert conditions. Despite the article's somewhat sensationalist tone, it represented a serious intervention, going a step further than many other works of the time in considering possible causes for the climate changes discussed. While Kropotkin did not fully endorse any particular theory, he underscored Svante Arrhenius's research on the effects of variations in atmospheric carbon dioxide and cited changes in volcanic activity as a possible driver of climatic variations.[41]

Kropotkin's contemporaries, however, remained unconvinced by Arrhenius's research. In fact, Arrhenius and his proposed mechanism of climatic changes through changes in the atmospheric makeup lingered in the margins of the debate. None of the proposed causes of climatic changes seemed fully convincing to the climatological community at the turn of the century. And this lack of a consensus on what could cause the hypothesized large, or even global, climate-driven desiccation would become a sore spot for practitioners.

A Dead End?

In 1901, three years before Kropotkin's article, the Swedish meteorologist Nils Ekholm, a colleague and friend of Arrhenius, reviewed the debate on large-scale climate changes. He expressed his doubts about drawing any definite conclusions, as quantitative meteorological data series did not reach far enough into the past. Citing the available data from various European cities, Ekholm concluded that it was impossible to determine whether the climatic variations during the past hundred fifty years were periodic, progressive, or accidental. This was certainly a sobering note for climate scientists. Nine years later, the summary of a Stockholm conference on climate changes in the Late Quaternary reflected this rather defeated mood among climatologists and conceded that "we apparently still stand at the very beginning of research."[42]

Almost half a century after the debate had begun, climatologists were as divided as ever about the existence of large climatic shifts in historical times, without a clear path toward a shared understanding. This was partly because the scientists could not agree on common methods and standards of evidence—a problem that the German geographer and historian Joseph Partsch had already addressed in 1889 but that was still unresolved on the eve of the First World War. In his article, Partsch criticized the "uncertainties of method" among climate scientists, who all too often relied on imprecise and exaggerated information. He conceded that the dearth of meteorological data from former centuries and the lack of means to determine the exact timing of geological changes made historical methods indispensable for climate scientists. The problems that Partsch saw with historical sources and eyewitness accounts, however, were the all-too-human flaws of emphasizing weather extremes and confusing correlation with causation.[43]

But what could climate scientists do to gather more accurate climatic information about the more distant past of human life on earth? Partsch himself advocated further studies on the historic water levels of endorheic lakes, which Humboldt had already used for his own climate studies in South America. Compared to the interconnectedness of other bodies of water, endorheic lakes had the marked advantage of representing bounded and less complex watersheds, which made historic fluctuations of the water level more significant for climate studies. Heeding Partsch's call or their own doubts about the available data, other climate scientists set about mining historical texts for quantitative—or at least quantifiable—data on water levels and precipitation. Maybe the most astounding result of this search was the information supplied by Hermann Vogelstein, a rabbi and scholar of old Hebrew texts. In his study on agricultural practices in Palestine during the time of the Mishnah, he referred to ancient recordings of precipitation volumes taken with standardized clay receptacles. Vogelstein converted the recordings into metric measurements, which climate scientists then compared with biblical accounts and contemporary meteorological data from the same region. Other practitioners turned to biological sources of evidence for recent climate changes, such as the study of tree rings.[44]

Despite these ingenious attempts to gather quantitative climatic data from past centuries, reliable evidence about past conditions remained spotty at best. And thus the climate change debate resulted in a rather paradoxical situation around the turn of the century. With the general acceptance of the glacial theory and the evidence of large environmental transformations in the past, the notoriously divided community of climate scientists appeared to have come to something resembling a consensus that the climate was not inherently stable. This consensus, and the associated discussion about climate variability as either progressive or cyclical, initially paved the way for a wealth of new studies on climatic pulsations and changes. The renewed scientific interest was then cut short, however, by the continuing, and increasingly glaring, lack of any convincing explanatory mechanisms for climatic shifts independent of anthropogenic actions. Ultimately, this shortcoming had the power to undermine the integrity and reputation of the still young field of climatology altogether, leading practitioners to reevaluate geographical and geological methods, which by the beginning of the twentieth century were visibly out of tune with the increasingly sharp distinctions between the sciences of nature and culture.[45]

Many of the studies on climatological variability—whether progressive or cyclical—had in common a sudden uncertainty and evasiveness when it came to discussing the causes of these changes. Practitioners advanced competing hypotheses about solar influences, shifts in the shape of the Earth's orbit through gravitational pull (proposed by James Croll in the 1870s), other changes in the Earth's axis alignment, and the impact of volcanic activity.[46] But most annoyingly for climate scientists, none of these explanations was verifiable. Brückner, for all of his detailed work in gathering meteorological and climatological data from around the world, became extremely vague about the mechanisms behind his proposed thirty-five-year cycles of climatic variation. In the end, his short treatment of the subject culminated in an indistinct sense that the sun had something to do with it.[47]

Faced with this causal uncertainty, many practitioners abandoned the research into large-scale climatic changes altogether. Supporters of Theobald Fischer and his hypothesis of a "natural" desiccation had an increasingly difficult time finding a receptive audience among their peers. Without any hard evidence on the causes for autogenic climate changes, investigations were bound to end in conjecture. By the eve of the First World War, climatologists were no closer to a solution, and the debate seemed to be going in circles. Kropotkin's warnings of a large and inexorable desiccation of Eurasia were still discussed in European journals but were now increasingly dismissed as uncorroborated and unverifiable. There was now a clear trend to attribute climatic changes to local human action. And even climate scientists who insisted on the existence of larger postglacial climate changes became increasingly unwilling to locate them in specific time frames.[48]

After carefully reviewing the extant literature, the Austrian geographer Hermann Leiter concluded in 1909 that the frequently asserted desiccation of North Africa could not be proven. Five years later, the British geologist and explorer John Gregory expanded Leiter's conclusion to apply to the whole world. Despite his reservations about a fully conclusive answer, he ultimately replied to his own article's question "Is the Earth Drying Up?" in the negative, asserting that while there had been large climatic changes in recent geologic times, there had not been any during historic times. Of course this was a problematic claim as well: if telluric or cosmic processes had been at work in the deep past, why had they ceased to be active in the past ten thousand years? What had happened to the powerful forces that had brought about ice ages and arid periods in the deep past of the earth?[49]

The climate change debate seemed to have hit a dead end. Climate scientists could not explain why climate changes independent of human action occurred (or why they had *not* occurred since the last ice age), nor did they have sufficient data to back up their various claims about what had actually happened to climatic conditions during historical times.[50] It was not until 1924 that the Serbian polymath Milutin Milankovitch published his findings that the earth's orbital variations could cause large climatic changes, and it took even longer for the theory to become widely accepted. Similarly, Arrhenius's hypotheses about an anthropogenic amplification of the atmospheric greenhouse effect would go largely unnoticed until they gained traction in the second half of the twentieth century. Despite new information on historic lake levels in the Sahara and longer data series on precipitation, the evidence for climatic changes also remained spotty and impressionistic. Scientists increasingly attacked eyewitness accounts, whether from antiquity or from recent scientific explorations, for their alleged inaccuracy and lack of objectivity, and the newly standardized meteorological data series did not reach far enough into the past to permit any definite conclusions. Even the geographical knowledge of desert regions remained incomplete until motorized and aerial exploration began in the 1920s and 1930s.[51]

Moreover, larger intellectual processes were at work that tipped the scale even further against the postulation and acceptance of climatic variability, let alone climatic changes: particularly in the German-speaking parts of Europe, increasingly popular theories of race harkened back to earlier ideas of climatic stability, as they often explicitly assumed the stability of climatic conditions. Environmental variation was inconvenient for the conception of essential racial traits honed in permanent and stable, if regionally different, surroundings. This formed the foundation of Willy Hellpach's theory of climatic influence on racial psychology that became popular in Germany in the early twentieth century.[52]

To be sure, race-based climate theories were common in other parts of the world as well and had indeed been a staple of the nineteenth century. And theories of race did not inherently presuppose stable climates. In the United States, the prolific eugenicist Ellsworth Huntington developed an environmentally deterministic theory of history and cultural development that explicitly posited climatic variations and their impact on human societies and racial evolution. In his research on climate, Huntington was as attracted to the dry regions of the Mediterranean basin and Central Asia as his European predecessors had been. But his sprawling argument and his attempts at a holistic explanation of world history did not convince most of his colleagues in the Old World. After the First World War, Partsch—still as critical of methods in climatology as he had been thirty years before—would write a scathing critique of Huntington's work on the Middle East, referring to its author with the underhanded epithet of a "spirited [*geistvoll*] geographer."[53]

Changing Climates

Despite all of the issues and uncertainties, the climate change debate never fully vanished in the first half of the twentieth century. While the First World War and its aftermath interrupted international exchanges among European scholars, questions about changing climates cropped up in academic journals once again in the 1920s, just as international connections were being rebuilt. And while the European climatological community did not take Huntington too seriously, he found a receptive popular audience. In fact, his ideas further fueled the already widespread anxieties about large environmental and climatic changes that had been inspired by half a century of discussions about ice ages, desiccation, and climatic instability.[54]

If, as Richard H. Grove and Vinita Damodaran have argued, the 1860s had been the "first environmental decade," in the fifty years that followed public environmental sensibilities were honed not only on the increasingly popular theories of evolution but also on ideas about climatic instability. By the early decades of the twentieth century, notions about changing climates and the associated ideas of environmental decline and catastrophe had long made it out of the ivory tower and had become a popular topic of discussion. While climatology as a discipline fell victim to its own early success, with different factions among the growing number of practitioners fighting over the definition of the field, preferred methodology, and elective affinities with either the telluric or atmospheric sciences, popular ideas of climatic change had their heyday.[55]

In fact, already in the nineteenth century notions of climatic change had inspired discussions about how "declined" environments with extreme climatic conditions could be resurrected and rendered useful. Johannes Walther, who became known as Wüstenwalther (desert Walther) in the German geological community, bridged the scientific description of desert environments and their potential use in his work. While Walther was mainly concerned with desert formation—itself an issue that arose out of the climate change debate—he used some lines of his foundational work on the "law of desert formation" to ruminate about the future of desert environments around the world: "Our developing culture will press on and on towards the deserts, to desalinate the alluvial soil and make it productive through artificial irrigation." Some of Walther's contemporaries and even some of his predecessors were convinced that the time to exploit the deserts had already arrived. They speculated on the directed modification of the North African climate, which would potentially not only transform the Sahara but also—through atmospheric effects—positively affect landscapes as far afield as the deserts of the Middle East and even the steppes of Central Asia.[56]

The projects that these climate modification advocates had in mind had first emerged on the coattails of the climate change debate in the second half of the nineteenth century. And in fact the debate about climate change was always closely connected to debates about actively changing climates. This was apparent in the notion of anthropogenic climate change and desiccation through deforestation in North Africa, which, as the common argument went, could be reversed by planting trees and restoring old irrigation systems. But even beyond these ideas of covering the arid landscapes of North Africa in forests, some enterprising individuals thought about other, more technology-intensive schemes to transform desert environments and climates. Theobald Fischer was not the only climate scientist to voice the belief that desertified environments could be reclaimed by humans through the "means of modern technology."[57]

It is not clear whether Fischer was thinking of the French engineer François Roudaire in this context. Other participants in the climate debate, however, did directly refer to him or to the preparations for his project.[58] Taking the ideas that climates in North Africa *had* changed and *could be* changed to their extremes, the enterprising Roudaire devised an ambitious plan to rebuild the "granary of Rome" using the full force of modern technology at his disposal. His Sahara Sea project, which is the subject of the next two chapters, was designed to attenuate the climate in a particularly hot and arid region of the Tunisian Sahara.

2

Flooding the Desert

ROUDAIRE'S SAHARA SEA PROJECT

HENRIK IBSEN'S *Peer Gynt* premiered in Christiania, Oslo, in 1876.[1] The audience was treated to a refreshingly unorthodox plot. Ibsen's protagonist, a boisterous ne'er-do-well, tumbles from one outrageous adventure to the next. He hunts reindeer (or talks about it at least), abducts a bride on the night before her wedding, drinks to inebriation with trolls in the Scandinavian mountains, and—in the play's fourth act—ends up on the shores of North Africa. Besides fighting a Greek revolt, being hailed as the prophet of a local tribe, and setting out to wander Egypt as a self-made historian, the hero also finds himself in the Moroccan Sahara, deliberating the purpose of "this patch of the world, that for ever lies fallow." Peer starts to ruminate, looks over to the Mediterranean, and sketches a plan:

> Is that sea in the east there, that dazzling expanse
> all gleaming? It can't be; 'tis but a mirage.
> The sea's to the west; it lies piled up behind me,
> dammed out from the desert by a sloping ridge.

And then, Peer is suddenly struck by an idea:

> Dammed out? Then I could—? The ridge is narrow.
> Dammed out? It wants but a gap, a canal,—
> like a flood of life would the waters rush
> in through the channel, and fill the desert!
> Soon would the whole of yon red-hot grave
> spread forth, a breezy and rippling sea.
> The oases would rise in the midst, like islands;
> Atlas would tower in green cliffs on the north;
> sailing-ships would, like stray birds on the wing,
> skim to the south, on the caravans' track.

Life-giving breezes would scatter the choking
vapours, and dew would distil from the clouds.
People would build themselves town on town,
and grass would grow green round the swaying palm-trees.
The southland, behind the Sahara's all,
would make a new seaboard for civilization.[2]

Peer jumps up, full of inspiration to create "Gyntiana," a new kingdom in a transformed and fertile Sahara. Alas, the grand plan remains but a short-lived obsession. Distracted by new adventures in other parts of the world, Peer abandons his scheme as quickly as he came up with it.

Almost forty years later, a strikingly similar project progressed far beyond the planning stage in a novel by the French adventure and science fiction writer Jules Verne. In his voluminous oeuvre, Verne had his protagonists ride balloons around the globe, undertake submarine voyages to the depth of the ocean, and voyage to the center of the earth. In the last of his books published during his lifetime, Verne turned his attention to the Sahara. Published in 1905, *The Invasion of the Sea* tells the story of the engineer and entrepreneur de Schaller, who studies the possibility of flooding a low-lying area in southern Tunisia with water from the Mediterranean. While the local Tuareg stage a revolt and capture Schaller and his military escort, a natural catastrophe intervenes: a deus-ex-machina earthquake pierces the rocky barrier along the coast and floods the large desert depression with water from the Mediterranean, preempting de Schaller's engineering project. Amid large-scale destruction— the Europeans survive while the insurgent Tuareg perish in the tsunami—a new inland sea is formed in the desert.[3]

Like most of Verne's novels, *The Invasion of the Sea* does not stray too far from the places, events, and technologies of the author's time. It was, in fact, based on the *mer intérieure*, or Sahara Sea, a very real project that was the subject of much scientific and public debate in France of the 1870s and early 1880s. De Schaller is Verne's rather feebly disguised fictional version of the famous French engineer Ferdinand de Lesseps, the developer of the Suez Canal and one of the staunchest defenders of the Sahara Sea scheme, which was developed by his friend, colleague, and compatriot François Roudaire. Verne's geography of southern Tunisia is accurate, and he refers explicitly to the soundings and surveys that were undertaken in the region in the second half of the nineteenth century. Only the earthquake is pure literary imagination. Similarly, Ibsen probably did not invent Gyntiana out of thin air. He had most likely picked up on the first designs to flood low-lying areas in the Sahara, which started to appear in geographical journals and the popular press in the 1850s and 1860s. So rather than the "poet anticipating the engineer," as a 1928 article on man-made inland seas in the Sahara put it, one of the most astounding engineering projects of the second half of the nineteenth century had actually inspired the poetic imagination.[4]

Roudaire's Sahara Sea would have flooded a number of low-lying desert depressions in southern Tunisia and Algeria with water from the Mediterranean—very much akin to de Schaller's and Peer Gynt's visions. In Roudaire's first calculations, the resulting man-made lake would have covered an area of about 16,000 to 19,000 square kilometers (or about 6,200 to 7,300 square miles; somewhere between the size of Connecticut and New Jersey). The French engineer expected easier and safer access and new marine trade routes to the inland areas of French Algeria, but also a climatic transformation that would turn the desert into fertile land for agriculture and European settlement. Behind the idea stood assumptions of a once fertile North Africa and of the variability and malleability of climatic conditions through geological or human action, which had emerged from the contemporaneous climate change debate.[5]

Although all but forgotten today, the Sahara Sea project was a sensation in its time. It inspired not only an enduring international debate but also a long line of evolutions and imitations reaching well into the twentieth century. While Roudaire and his successors would never enjoy the sight of a flooded Sahara, their project designs represented the most ambitious expression of new ideas about environmental transformation that had been on the rise since the mid-nineteenth century. The completion of the Suez Canal in 1869, in particular, inspired engineers—and particularly colonial engineers—to envisage further large-scale modifications of geography. Once the development of climatology and new theories of desertification added to the proven capacity of modern technology, the concepts and tools were in place for turning dreams of engineered climate change into tangible and more or less quantifiable projects on an unprecedented scale.

The Sahara Sea was one of the first projects that today might be classified as "geoengineering." In contrast to the trans-Saharan railway project that was discussed around the same time, Roudaire's plans aimed at transforming—and not just bridging—the space, climates, and environments of the Sahara. In debates about glaciation, ice ages, and climatic shifts, the earth had lost its aura of immutability. This opened the door for colonial engineers who had ambitions to ameliorate nature and saw in the Sahara the perfect testing ground for their designs: a frontier of colonial knowledge and control and a vast, empty space in dire need of improvement. The history of the Sahara Sea begins in this desert, or more specifically in the arid interior of the French colony of Algeria. Then, the plot quickly spirals outward—to France, Tunisia, Morocco, Great Britain, Germany, and beyond.

Colonized Sand

In an 1887 treatise on Africa, the French geographer Lucien Lanier recalled an almost fifty-year-old report that passed a crushing judgment on the colonial potential of Algeria: "One finds in the colony neither stones, nor water, nor forest," it read. Lanier was a bit more optimistic in his own appraisal: there was clearly more than just sand in the colony, but he too highlighted the scarcity of vegetation and water. Reproducing a common colonial argument, Lanier believed that in the recent past lush forests had covered the country but had recently been degraded by careless indigenous practices of burning and pasturage. In contrast to many of his contemporaries, Lanier also put some blame on the French colonizers and their own land practices. Still, Lanier was largely wrong: while the woodlands would shrink even more under French rule, forests had barely covered 2 percent of Algeria before colonization. Except for a narrow fertile band on the Mediterranean coast north of the Tell Atlas, Algeria was and remains a semiarid to arid country in which water resources are generally sparse and very unevenly distributed over the territory and the seasons. In many parts, both seasonal water shortages and multiannual or even decadal drought periods seriously restrict human activities.[6]

In the early years of French Algeria, the would-be colonizers mainly framed the environmental conditions of the territory in terms of their effect on military operations, as colonial troops were busy trying to suppress strong and persistent armed resistance to their early incursions into the interior. To Europeans, the Sahara remained a mysterious vast entity beyond the scope of their colonization efforts: a "country of thirst" that starred mostly in the often inaccurate—or at least overly generalized—representations of Algeria in the metropole. Although their continued attempts to establish military control over the entire colonial territory soon brought the French military into direct contact with the desert and its inhabitants, the French government's first, rather feeble, attempts at colonial development and settlement focused almost exclusively on the coastal regions of Algeria. Here, colonization took the form of "centers"—areas that the army occupied by force before being settled by French civilians.[7]

Starting in 1849, French engineers constructed the first dammed reservoirs in Algeria, but building larger irrigation works posed more problems than the colonial government had anticipated. There were a number of dam failures, such as the widely reported ones at Fergoug in 1881 and Cheurfas in 1885. European engineers were not accustomed to Algerian soils and were unacquainted with *wadis*, riverbeds that are dry for most of the year and then fill up rapidly during or shortly after rainfalls, posing structural challenges to dams designed for European environments. Rebuilding broken dams, clearing silt, and reinforcing existing irrigation structures only exacerbated the already high costs of maintaining the colony. Although the colonial government tried to reduce its expenditures by abandoning its policy of full financial support for water supply measures in French settlement centers from 1870 onward, the costs were still steeply on the rise in the 1870s.[8]

Another factor that drove up colonial expenses and added to the French sense of colonial crisis was the enduring volatile security situation, as the colonized Algerians continued to resist the occupying French forces. Uneasy about the enduring threat of unrest, Europeans settled in cities and not—as anticipated by official plans—in the countryside and further away from the coast. By World War I, four hundred thousand Europeans were working as farmers, a substantial number that still fell well short of official goals. Even the "patriotic emigration" from Alsace and Lorraine after the French loss against Prussia in 1870 had not helped the situation: indeed, the success of the program that the French government touted was mostly a myth, as more people had moved from these regions to Algeria before the Franco-Prussian War than after.[9]

And yet the perpetuation of the Alsatian myth among officials hints at the renewed prominence of Algeria in the French public sphere after 1870. During the Third Republic, the idea of using colonized land to make up for the upsetting territorial losses France had sustained on its eastern border gained great currency. Politicians advocated for populating Algeria with European freeholders. While that mostly failed in practice, it created a space for debates on colonial development. The demographic numbers coming in from North Africa were worrying French officials: the European population in Algeria grew much more slowly than the indigenous one, which recovered from a period of famine and epidemics and increased steadily between 1872 and 1886. With this development, French settlers and administrators became increasingly anxious over diminishing colonizable reserves, with the allegedly empty settler land filling up with a rising number of North Africans. Colonial officials attempted to counteract this purported threat by bringing Algerian-held land under direct colonial or individual European control. To this end, the colonizers employed dubiously legal purchases of property and straight expropriations and theft of communally held Algerian land.[10]

These land grabs were not the only methods that the French used to claim property. Their efforts also included large projects of land improvement, in particular forest conservation and afforestation measures. The colonial justifications that accompanied these attempts—the fight against an allegedly advancing desert, the attempts to rebuild the "granary of Rome," and the blame of indigenous practices for forest destruction and desertification—were closely tied to the property theft. In a classic story of European imperialism, the French used the purported need to conserve and protect nature as a pretext to remove land from Algerian control and force the indigenous population into a market-driven colonial economy. In fact, the activities of the Algerian Forest Service to bring the limited wood reserves under its purview often diametrically opposed the interests of Algerian peasants and herders, who relied on forest resources for their economic activities or even their very survival. Ironically, deforestation in North Africa was actually at its highest level under French occupation between about 1880 and 1940 and was exacerbated by the French practice of pushing Algerians further into marginal lands on forested mountain slopes. And while a minor net deforestation might have occurred even before the French invasion, it was far less extensive than colonial officials believed at the end of the nineteenth century. The French diagnosis of a steadily declining North African environment since ancient times was, as in so many colonial cases, a misreading of the landscape.[11]

Ignoring these realities, the French propaganda machine was running at full speed. Associations such as the Algiers-based Reforestation League called for complete state control of all woodlands in Algeria and emphasized the need for quick action with vivid language: "Every deforested country is a country condemned to death!," one pamphlet read. The arguments had staying power and produced a great wealth of scientific and popular literature that often linked directly to the larger debate on climate changes in North Africa. If the problem was the ever worsening scarcity of forests, then the solution could only be to create more spaces where trees could grow. This required transforming Algeria's arid environment. It meant that, in one way or another, the colonial regime had to "improve" the desert to yield agricultural settler land.[12]

The Sahara Sea

The French engineer François Roudaire was not alone in falling prey to the draw of the desert. In his study of French violence in the Sahara, Benjamin Claude Brower has described the phenomenon as the "Saharan sublime," a simultaneous attraction and revulsion felt by the French colonizers when confronted with the grandiose and barren desert landscapes of North Africa. To European minds, the sand dunes and vast rocky planes appeared as remnants of an apocalyptic event that had left the landscape devastated. Roudaire, as well, expressed a simultaneous fascination with and aversion to the desert, channeling this response into a project that claimed to revert the past apocalypse and restore the fallen land to a pristine and peaceful environment. Once he had conceived its basic outlines, the Sahara Sea became the overwhelming focus of his work and life.[13]

FIGURE 2.1. François Roudaire in Algeria (1879). *Source*: Public domain, but originally from the "Collection personnelle Bertrand Bouret."

François Élie Roudaire was born in 1836 in Guéret, a commune in the Limousin region of central France. He excelled in his studies at the local secondary school and, at eighteen, entered the French military academy Saint-Cyr. After completing his studies, Roudaire was among the few students admitted to further courses at the École d'état-major. Upon graduation, and with a commendation for his excellence in topography, Roudaire immediately joined the geographical service of the French army in Algeria. He would stay in the colony for seventeen years, helping to survey many parts of the large territory. When the Franco-Prussian War erupted in 1870, Roudaire was called back to France to help defend the country. He was wounded by a gunshot in the Battle of Werth and recovered quickly at a military hospital. After the French defeat, Roudaire returned to Algeria to continue his work with the geographical service but soon started to travel back and forth between the colony and Paris to campaign for his Sahara Sea.

The project focused on a region in southern Tunisia and Algeria that Roudaire had become familiar with on one of his surveying trips. The whole area, stretching around two hundred miles west from the Gulf of Gabès, is an arid to hyperarid landscape known as the region of the chotts (Figures 2.2 and 2.3).[14] Also known as *shebkas*, the chotts are low-lying salt-covered depressions ten to thirty miles wide that every once in a while fill up with runoff water from heavy desert downpours. In the words of Roudaire, the chotts resembled "vast plains covered with snow or white frost." In a different article, the French engineer also mentioned another illusion that is created when sunlight reflects off the white ground, producing the semblance of an enormous lake filled with water. That, in fact, was exactly what he hoped to create in reality with the Sahara Sea.

From his earliest days in the region, Roudaire had been speculating on the origins and history of the desert depressions. He certainly had done his research: in one of his longer publications, Roudaire included word-for-word descriptions of the area written by North African and Middle Eastern travelers. At the Academy of Sciences in Paris, Roudaire's friend Ferdinand de Lesseps would add to this list by citing a manuscript found in a mosque close to the

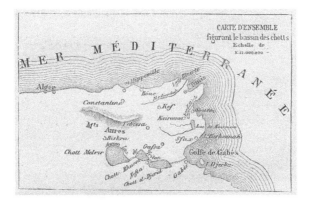

FIGURE 2.2. Map of northern Algeria and Tunisia, showing
the site of Roudaire's project and the proposed canals.
Source: P. H. Antichan, *La Tunisie. Son passé et son avenir*
(Paris: Delagrave, 1884), 290.

chotts. The text told the story of an ancient city named Zaafrane, a once fertile
and mighty settlement on the banks of a sea or large lake, which eventually
dried out, leaving a basin that became covered in crystallized salt. Now all that
was purportedly left of the erstwhile grandeur was a small desert settlement.
This manuscript also confirmed some of the classical accounts, which gave
Roudaire the idea that the chotts—or at least one of them—might once have
been known under a different name: Lake Tritonis, a perennial body of water
in lush and fertile surroundings.[15]

Roudaire based his theories first and foremost on the ancient Greek historian and geographer Herodotus, who around 450 BCE had portrayed Lake
Tritonis as being fed by a river of the same name, at the center of a prosperous
region somewhere in northern Africa. The lake had also been described by the
so-called Pseudo-Scylax and by the Roman geographer Pomponius Mela.[16]
Likewise, the Egyptian polymath Ptolemy had referred to a river of the same
name in his writings. The location of the presumed lake remains a mystery that
to this day has not been conclusively solved. Upon seeing the chotts, however,
Roudaire was immediately convinced that the depressions were nothing other
than the dried-out basin of the lost mythological lake.[17]

Roudaire had not come up with this idea on his own accord. He was familiar with the writings of the British traveler Thomas Shaw, who had already
speculated on the chotts as the last remnant of the lost Lake Tritonis in the
mid-eighteenth century. In 1800, in a large study on Herodotus, the British
historian and geographer James Rennell had concurred with his compatriot's findings. Even this early, however, the theory did not go unchallenged.

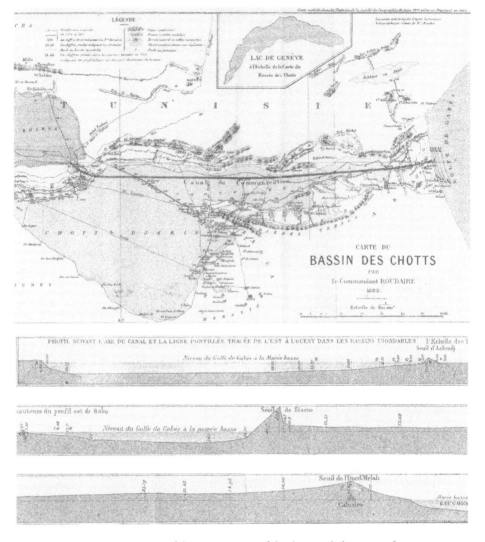

FIGURE 2.3. Map of the eastern region of the chotts with the proposed canals and new bodies of water (1883). *Source*: M. Gellerat, *Note sur la mer intérieure africaine ou Mer Roudaire* (Paris: P. Dubreuil, 1883).

The German Konrad Mannert, for example, disagreed in his mammoth ten-volume work on the "Geography of the Greeks and Romans" and identified Herodotus's Lake Tritonis as the Gulf of Gabès, whose semicircular coastline could have fooled the Greek historian into believing that it was separated fully from the Mediterranean Sea. Ignoring the dissenting voices, Roudaire remained convinced of Shaw's theory. It would, in fact, become a cornerstone of his argument for the engineered Sahara Sea, which he saw as restoration of an earlier, more perfect state of nature. Roudaire also explicitly referred to *re-creating* former environmental conditions to downplay the undeniably radical nature of his project and thus convince the more cautious among his readers. His emphasis on classical precedent also chimed with the widespread French rhetoric of the time that framed the French presence in North Africa as the continuation of Roman colonization efforts.[18]

Roudaire's plan was to connect the larger chotts with short canals before cutting through the bedrock sill, the *seuil* Melah, to introduce water from the Mediterranean into the Chott el Djerid, the depression closest to the Tunisian coastline. From there the water would spread to the other chotts, re-creating Herodotus's lake as the new *mer intérieure*. Once again, Roudaire was building on precedent: ten years before Roudaire's first publication in 1874, Charles Martins and Henri Duveyrier had both floated the possibility of creating an inland sea in the region. The latter, the famous Saint-Simonian explorer and a student of Heinrich Barth, would later become one of Roudaire's staunchest allies. In 1867, another article—this one written by Georges Lavigne—discussed the feasibility of inundating the chotts. "The Sahara—that's the enemy," Lavigne wrote programmatically, as he called on engineers to undertake detailed studies of the area.[19]

Lavigne could already rely on the first barometric altitude measurements of the area, which a mining engineer had taken in the 1840s. The recordings had revealed that the chotts were at least partly below sea level, but some critics continued to express doubts over the accuracy of the early altitude measurements. In his own measurements conducted from 1874 to 1875, Roudaire attempted to set the record straight. With support from Paul Bert, a physiologist and politician, Roudaire had secured ten thousand francs from the National Assembly for his expedition—a considerable sum that was further augmented by a subsidy of three thousand francs from the Society of Geography in Paris. Yet the money was barely sufficient for the large expedition that left France: Roudaire set out in the company of a number of scientists, protected by an escort of more than fifty soldiers.[20]

Not everything went smoothly on the trip: all three mercury barometers that the expedition had brought with them broke within the first month, which left Roudaire with only five aneroid barometers. To make matters worse, fevers ravaged the expedition, and at least one member fell so seriously ill that he had to be hospitalized in Biskra. As financial constraints had forced Roudaire to abandon the idea of traveling with a water supply column, the expedition had to stay close to wells and could not venture out deeper into the chotts. On his return to Paris, Roudaire could nonetheless confirm that the Chotts Melrir and Sellem (a small chott east of the Chott Melrir) and a part of the Chott Rharsa were on average between twenty and twenty-seven meters below sea level, respectively. The engineer also reported that in the course of the future submersion of the chotts, three small oases and their settlements would probably be inundated. The project of the Sahara Sea was clearly under way.[21]

Roudaire presented the results of his trip at the 1875 Congrès international des sciences géographiques, which ultimately voted to support another mission for further and more accurate measurements of the Tunisian part of the chotts. Commissioned by the French Ministry for Public Instruction, Roudaire returned to North Africa the following year, accompanied once again by several French scientists. The group measured the altitude of the Chott el Djerid in Tunisia, which to Roudaire's chagrin proved to be largely above sea level. Roudaire, however, did not let that fact stifle his enthusiasm for very long. He remained convinced that his theory of Lake Tritonis was correct and that his plans to flood the chotts would still be feasible in a modified form and with the help of some dredging work. Roudaire's 1877 report for the French government was as optimistic as ever. And his confidence seemed to be contagious: the report received an overall positive response, and copies of it were sent out to prominent politicians, scientists, and public figures—among them Jules Verne, who thus owned the perfect reference work when he started work on his *Invasion of the Sea*. Plans of Roudaire's project were even exhibited at the Paris World Exposition of 1878, where they reached a larger international audience and enjoyed a generally encouraging reception.[22]

This marked the high point of success for Roudaire. He had managed to not only get his government's attention but also spread his ideas to a receptive French public. In the 1880s, the debates about the project continued, with arguments for and against its feasibility, but governmental funding dried up. The results from Roudaire's third mission to the chotts in 1879 were partly to blame for the waning enthusiasm: new measurements contradicted his earlier claims about the possible existence of an inland sea in historic times. Nevertheless, Roudaire continued to promote his ideas and in 1881 published a long report of his latest findings. By and large, the report did not add anything new to the project but complicated it further by taking into account the increasingly discouraging findings about topography and the geological condition of the bedrock sill. The areas of the chotts were smaller than Roudaire originally assumed, the depressions themselves had a higher elevation than he had hoped, and the connection between the chotts and the Mediterranean turned out to be made of both rocky and sandy soil strata, complicating the construction and eventual maintenance of the projected feeder canal. While public support started to wane, Roudaire seemed unfazed: he believed the expected gains of the completed project to be so large that he felt permitted to disregard the skyrocketing costs of the Sahara Sea.[23]

Engineered Climate Change

Roudaire and his fellow campaigners claimed that the Sahara Sea would provide a host of benefits, first of all by adding to France's glory and reputation with a magnificent and unique engineering work. This argument fell on fertile soil in the years after the shock of defeat against the Prussian armies, when the French public, as well as government officials, welcomed anything that promised to restore some of France's lost grandeur with nationalist vigor. Roudaire also argued that the project would help solidify France's colonial authority in Algeria. In his first major report, he imagined North Africans being awestruck and full of reverence at the sight of the Sahara Sea. Touching on military matters, Roudaire highlighted the role of the man-made body of water in making the hinterland of Tunisia and Algeria more accessible to the French army, easing the conquest and control of the people who lived there. Roudaire also foresaw great marine traffic in the flooded chotts and an active economic life around the new coastline. One of the ports that would be established on the banks of the man-made lake, he argued, could become the central trading center for the whole of North Africa, shortening the necessary caravan trek through the desert and wresting control of Saharan trading routes from the hands of the indigenous population.[24]

In his writings, Roudaire also added a rather unexpected advantage of the interior sea over the trans-Saharan railway project. While inundating the chotts would surely destroy some oases, Roudaire argued, not many settlements were in the region, whereas laying miles upon miles of track for the railway project would force the French colonial government to expropriate more land all around the colony. More compelling for the colonizers, however, was a second advantage of the Sahara Sea over the railroad, which Roudaire cited tirelessly: the large body of water would act as a protective barrier that would not only impede destructive locust migrations but also halt the advancing sands of the Sahara.[25]

But what was it about the sands of the Sahara that made them seem so threatening as to merit a mention alongside the more conventional trade and military issues? Roudaire and his supporters were attuned to the continuing academic debate about the variability of the North African environment, and they picked up on the contemporary concerns about changing climates. Roudaire adopted arguments that North Africa was progressively desiccating and used them for his own ends. The *mer intérieure* was designed to do much more than simply block the encroaching sands of the Sahara. In fact, it represented one of the first large-scale engineering projects that explicitly chose climate as the target. Roudaire drew on the meteorological work of Antoine and Edmond Becquerel, who emphasized the role bodies of water played in moderating climates. Increased evaporation from the surface of the new bodies of water would lead to increased moisture in the air, and that in turn would lead to more rain "capable of fertilizing the desert, irrigating a sterile soil and multiplying the number of oases," as Lanier stated with unbridled optimism. Propositions of artificially altering climates were popular in Europe in the nineteenth century, but the combined scale, publicity, and serious consideration of the Sahara Sea project made it stand out. Roudaire claimed that the sea would have an evaporation rate of twenty-eight million cubic meters of water per day. And this calculation may have been on the lower end of the spectrum, as Roudaire was faced with a conflict of interest in having to keep the expected amount of evaporation reasonably low to make the maintenance of the inundated chotts through the inflow of water from the Mediterranean seem possible.[26]

The effect of the evaporation would, in any case, be substantial, Roudaire argued: he projected that the evaporated moisture would form clouds that would release their moisture as rain at the foot of the Aurès Mountains further north of the chotts. Without reliable data on wind patterns, Roudaire backed up his claims with a comparison to the *lacs Amers*, or Bitter Lakes, which had been constructed in the course of the Suez Canal works. As he had learned from his fellow campaigner de Lesseps, the man-made lakes had a noticeable influence on the local climate: vegetation had sprung up around the banks and precipitation had increased markedly. Roudaire argued that the Sahara Sea would increase the evaporation surface of this model and thus also scale up the climate-changing effect. He also hoped that the Sahara Sea would moisten the dry sirocco winds from the Sahara, which climatologists believed to have a damaging effect on agriculture in southern Europe. All of this together would produce altered climatic conditions on an unprecedented scale: Roudaire anticipated that the whole region of the chotts would be transformed into "an immense oasis of 600,000 hectares [about 1.5 million acres]." In his calculations, the engineering project would in fact do away with part of the Sahara itself.[27]

The scale of the project was undoubtedly grand. To completely fill the chotts through the projected canal would take not days or months but several years. But, Roudaire contended, all of the effort and time would be well spent to wrench a piece of land from the desert's grip. In his 1883 publication, he went into more detail about the mechanism of climate change, which he had adopted from John Tyndall: since water vapor has at the same time a great transparency for light and a great opacity for heat (infrared light), the evaporated water would act as a "protective shield," absorbing the heat from solar radiation before it hit the earth's surface during the day and reducing heat loss at night by absorbing infrared radiation from the earth's surface. This effect would moderate the climate in the region by reducing the desert region's stark differences in temperature between day and night, making the area around the chotts inhabitable and possibly even fertile.[28]

The argument certainly meshed well with French attempts to make Algeria not simply into a coastal settler colony but into an extension or mirror image of France itself. For French colonial officials, the Sahara, an impediment to both control and agriculture, stood in the way of this project. Even more worryingly, the Sahara actually seemed to be growing, according to the increasing voices warning of a creeping desertification of North Africa since Roman days. Roudaire picked up the widespread rhetoric about the granary of Rome in his own writings, arguing that the south of Algeria and Tunisia were incomparably more fertile during classical times. This, Roudaire argued in some of his writings, was because the chotts had dried out through some geological process or event—probably the slow silting up of a once existing strait. The prevalent contention among climatologists that local climate was fairly stable certainly helped Roudaire's cause in supporting his central claim that refilling the chotts was all that was required to refertilize the entire area. With this "simple" measure, France could convert a part of North Africa to a fertile region with a moderate climate, creating a kind of "second France." Modern hydraulic engineering made it possible for Roudaire and his supporters to reconceptualize colonialism not just as military occupation or as a vague civilizing mission but as the active transformation of landscapes and climates to meet European demands. The Sahara Sea project was part of this shift in emphasis that redirected the colonial focus from the anticipated Europeanization of the indigenous population to that of indigenous landscapes. Roudaire was the archetype of the engineer of High Imperialism who believed that industrial technology gave him not only the tools but also the mandate to transform a deficient nature.[29]

Like the academic discussions about desertification and climate change of the time, Roudaire's engineering project reveals the close connections between European colonial projects and nineteenth-century ideas and theories about the environment. The story of Roudaire's plans and their reception shows how—and how quickly—newly acquired knowledge of the Sahara and the concurrent discussions on climate change became entangled in debates about exploiting and technologically transforming landscapes in North Africa. Alongside plantation economies, colonial violence, and strategies of divide and rule, the serious consideration of large water engineering projects represented one of the defining characteristics of High Imperialism in Africa. Colonizers' enthusiasm for attempts to transform environments often concealed their continuing ignorance of the regions, especially in relation to the Sahara. By 1870, the number of European scientists who had traveled some distance beyond the borders of the Sahara and made it back alive was still in the low double digits. Yet the continuing dearth of information was covered up by a mounting belief in the almost limitless possibilities that modern industrial-age technologies promised. This optimism in technological progress prepared the ground for visions of colonial expansion and control, even in places that still appeared as blank white spaces on European maps. Ultimately, Roudaire's project would not fill these spaces. The Sahara Sea remained a project on paper. And while a whole host of issues sealed the fate of Roudaire's plans, fantasies of transforming colonial environments and climates on a large scale were there to stay.[30]

3

New Garden Edens

THE RISE OF COLONIAL
CLIMATE ENGINEERING

COMMENTING ON GERHARD ROHLFS'S SAHARA expeditions, the German cartographer August Petermann pondered the utility of deserts, citing the guano and saltpeter deposits in the Atacama in Chile and the successes in using the interior deserts of Australia for pasturage. "And so," Petermann reasoned, "the African deserts might one day also fulfill a now still unforeseen role for the world." What exactly that role would be was still up in the air in 1878, but the colonial belief that the Sahara could be entirely remodeled and made to serve an important economic function was widespread.[1]

Roudaire's project of transforming the Sahara entered the public debate at a time when European explorers had accomplished their first desert crossings and had started to distribute scattered firsthand information on the interior of North Africa. The European response to the Sahara quickly shifted from terror of the unknown to fascination, entrepreneurship, and entitlement. The long and serious consideration of Roudaire's Sahara Sea project is a prime example of how far the European belief in technology's power to change environmental and climatic conditions had progressed in the last third of the nineteenth century. This belief combined the optimism that resulted from the new and greater engineering expertise of the period with ideas of a variable and mutable nature—inspired not least by the debates on changing climates and growing deserts. If the climatic conditions of sizeable regions had been altered in historical time spans through either geological or inadvertent collective human actions, then the idea that similar changes could also be induced deliberately was not too far-fetched. In the heady times of imperial expansion, the prospect of converting the hostile desert into arable settler land had unprecedented appeal for Europeans, especially when it was embedded in the powerful rhetoric of hydroimperialism. The biggest problem, however, was the continued shortage of reliable data and information on geological, geographical, and climatological conditions.[2]

The French had made efforts to explore the Sahara as early as the 1840s under the energetic leadership of Ismayl Urbain, an interpreter who rose through the ranks of the colonial bureaucracy to become a member of the Commission for the Scientific Exploration of Algeria. Over the next few decades, European knowledge about the Sahara increased considerably, but the difficult terrain, along with the lack of infrastructure and the French dependence on often-reluctant indigenous guides and informants, meant that the data they collected remained piecemeal at best. In his influential 1845 study of the Sahara, Eugène Daumas had depicted a fascinating and varied landscape with ample opportunities for easy colonial development. His optimism was soon corrected by the reality of ongoing military conflicts with native populations and the hard-earned recognition that the arid environment in the interior posed a formidable challenge to European control. Not least because of the use of violence by the French army around the middle of the century, many inhabitants of the Sahara became loathed to cooperate with the colonial government, and the great North African desert remained largely outside the control and understanding of the French colonizers. It was not until the 1920s that aerial photography allowed for more accurate information and maps of the more remote parts of the desert. Similar issues plagued other disciplines as well, not least the field of climatology. Data on temperature and precipitation in the area of the chotts, so central to the core design of the Sahara Sea, remained sparse, with little to no reliable information that reached back further than a few decades.[3]

Despite this shortage of information, Roudaire's project inspired enthusiasm and a long-lasting, serious debate among engineers, archaeologists, geologists, and climatologists. Looking back at the discussions in 1887, Alphonse Rouire commented that the world of science seemed to have been "split in two" over Roudaire's scheme. This split was not, however, a simple matter of proponents versus detractors. The participants in the debate focused mostly on questions of technological feasibility and discussed the expected results of the project. The European adherents and critics overwhelmingly agreed that the Sahara had to be conquered; it was the suitability of the proposed means that they disagreed on passionately.[4]

Changing Deserts

To some critics, the Sahara Sea project appeared simply too expensive to be feasible. As his knowledge of the chotts' geology grew, Roudaire constantly had to revise his first optimistic calculations of the required budget. Moreover, how much of the region of the chotts could actually be flooded was never entirely clear, as critics questioned the precision and thoroughness of Roudaire's work. This is not too surprising, given that even in the early twentieth century a single barometric reading in the Sahara was still accurate only to about fifteen meters and involved a tedious comparison with synchronous pressure measurements at sea level. And regardless of their accuracy, the measurements gave only a few scattered data points covering but a small part of the chotts.[5]

One of the most persistent adversaries of Roudaire's plans was the geologist and mining engineer Edmond Fuchs. He had conducted his own surveys of the Tunisian part of the chotts in 1874 and had found the geological and geographical evidence at odds with Roudaire's plans. In surveying the barrier between the Mediterranean and the chotts, Fuchs estimated the highest point to be up to one hundred meters and the lowest points still at about fifty to sixty-five meters above sea level. This did not bode well for the canal that Roudaire envisioned in his plans for the Sahara Sea. Additionally, there was evidence that the barrier contained tertiary geological structures that pointed clearly to prehistoric—or, more accurately, Pliocene—origins of the bedrock sill. Taken together, this was quite a damning pronouncement on Roudaire's conviction that the chotts had been connected to the sea in historic times and would therefore be easy to reconnect. Even worse for Roudaire's credibility was the argument by some critics that through the constant and considerable evaporation of water in the desert region, the Sahara Sea would quickly salinize and eventually revert to its prior state: a chain of large salt flats.[6]

This still left the puzzle of Greek and Roman descriptions of the North African environment as rich and fertile. And on this point the debate about climate change became fully intertwined with the debate about Roudaire's climate engineering project. Fuchs had already argued in 1874 that the region's loss of fertility could be due to a long-term shift in the macro-climate or, as he put it, "a general cosmic phenomenon" that had desiccated North Africa. Referring explicitly to the unresolved arguments in the scientific discussions on large shifts in climate in North Africa, Alfred von Zittel likewise asserted that general meteorological events, not local geological ones, had turned the once water-rich and lush northern part of the continent into desert land. Zittel, however, saw no evidence for a great climatological change in *historical* times. Very selectively citing the climatological work of Theobald Fischer, he argued that the change to desert conditions close to the North African coast had most likely been caused by anthropogenic deforestation. This was the prevailing view among the European scientific community, but certainly not the only one. In front of the Academy of Sciences in Paris, for example, Gabriel Daubrée attacked Roudaire by arguing that the climate had in fact deteriorated around the

Mediterranean: the fertility that the Greek and Roman authors had described had noticeably declined, but this had not been the result of localized developments, as evidenced by similar and synchronous desiccation in Sicily, Spain, and Egypt.[7]

Zittel and Daubrée both considered the evidence from ancient texts and paired this approach with a close reading of recent findings in geology and the newly formalized sciences of meteorology and climatology. This was perhaps the most important shared feature of the very different visions and plans to reestablish North African fertility that accompanied the European colonial projects in the region. Dr. Paulin Trolard, one of the founders and the longtime director of the powerful Reforestation League in Algeria, argued in 1883 that North Africa could once again become the "granary of Rome." This colonial leitmotif was ever present in contemporary writings on Algeria. Sometimes it was even superimposed on the rhetoric Europeans used about the relative worth of the indigenous population, in which they posited Arabs as destructive and contrasted them with hardworking, proto-capitalist Kabyles or Berbers—whom Europeans thought of, maybe unsurprisingly, as the descendants of Roman settlers. This argument, Diana Davis has argued, amounted to a blatantly selective reading of the evidence, but one that was widely shared and accepted in colonial circles.[8]

While the vision of a once fertile North Africa became a trademark of colonial rhetoric, when the desertification had started and who or what should be held responsible for it remained obscure. One answer to the latter question was human action, but even then who exactly was to blame often remained unclear. François Trottier, who campaigned for reforestation in Algeria and came to be known as the "apostle of eucalyptus," shifted back and forth in his own writings. While he usually replicated and propagated the colonial narrative of indigenous and particularly Arab neglect of irrigation works and the environment, he also argued that the climate deterioration in Algeria was due to "thoughtless deforestations" that had occurred mostly *after* French conquest and had been committed by both Algerians and the French. Trottier also argued that precipitation had declined from 1855 onward. To what extent he truly believed this is difficult to assess. These claims certainly helped his cause and business of planting eucalyptus trees all over North Africa. This, by itself, was a large-scale environmental engineering project: Trottier hoped that the afforestation would engender a climatic modification, which he deemed necessary for the large-scale colonization of North Africa.[9]

Attempts to answer questions of climate change and climate variability in North Africa were bound to remain an exchange of impressions and preconceived notions for as long as the discussants lacked stable reference points. To evaluate potential shifts in climate, practitioners needed meteorological data over longer time periods. In 1864, the Société climatologique d'Alger was founded under the directorship of the medical doctor Émile Bertherand. Its findings were very much in line with colonial policy: the society deemed the climate of Algeria to be basically healthy for Europeans and was convinced that active climate modification in "unhealthy" areas was possible through irrigation measures, drainage, or afforestation. The Société, however, mainly focused on medical matters, and quantifiable information on the Algerian climate was still hard to come by. Only from 1874 onward did the newly founded Algerian Meteorological Service collect detailed data series on temperature, winds, and atmospheric pressure. This material would become vital for challenges to some of Roudaire's meteorological assertions. It showed, for instance, that the winds in the region of the chotts would usually carry humidity from the engineered sea not to semiarid northern Algeria but inland to an extremely dry part of the Sahara, where it would most likely be lost without any significant effect.[10]

This argument, among other criticisms, prompted the German geologist and Africa traveler Oskar Lenz to regard large-scale desert reclamation projects as short on sound scientific data and thus not worthy of serious consideration: in his blunt assessment, Roudaire's scheme was "futile" and other ideas of flooding the Sahara "too absurd to be discussed seriously." But were these ideas really that absurd? When Lenz discussed the origins of the Sahara, he implicitly assumed that there had been a time when the great desert had not been a desert at all. And while Lenz did not believe that water in North Africa had exercised a great effect on the climate, he did believe that the climate had changed dramatically over geological time scales. Similarly, Zittel criticized the idea of a submerged Sahara in historical times, while still believing that fresh water lakes had covered the desert in prehistoric times before drying out to form the salt marshes of the chotts.[11]

In the second half of the nineteenth century a geological worldview postulating a long history of the earth marked by colossal environmental and climatic changes emerged and became widely accepted. Once the past glaciation of Europe had become established theory, the debate did not stop. Geologists started to discuss the causes of its onset and end. Lenz attacked one of the explanatory models that held that a shift in the distribution of land and water on the earth's surface was responsible. Agassiz's student Éduard Desor, whose views on a flooded Sahara had inspired Roudaire, was one proponent of this theory. He argued that even relatively small changes in nature—such as the change of the water-to-land ratio in North Africa—could have large-scale effects. Thus, a renewed glaciation could happen even without a significant temperature decrease, if wind conditions changed. According to Desor's theory, the system of winds in Europe was controlled by conditions in North Africa and, in particular, by the Sahara, which he reverently labeled "the great regulator of our climate." Implicitly, this also meant that flooding parts of the Sahara anew could lead to large climatic shifts in Europe. Desor closed his argument with a dramatic appeal to keep the status quo: "May the Sahara remain a desert for a long time and, through its warm and dry breath, show the alpine glaciers their limits."[12]

With all their differences about the specific conditions in prehistoric North Africa, Lenz and Desor still agreed that the environment had undergone drastic changes in the past. And both also seemed to believe that drastic changes were still possible in the present—while disagreeing about the means and processes. Georges Lavigne, one of the early instigators of the debate on the chotts, clearly reflected the new uncertainty about a nature that now seemed to be in eternal flux: "Nature does not know any status quo," he wrote, "it has delivered Africa to two inimical elements; it has not said to the desert that it will always be a desert, nor to the Tell that it will always be the Tell; it has not provided permanent and insurmountable borders." The desert ceased to be an unchanging, dead landscape and became an alterable environment—a notion that helps to explain the emergence of great environmental transformation projects around the middle of the nineteenth century.[13]

The Climatic Machine

The Suez canal builder Ferdinand de Lesseps, Roudaire's staunchest supporter, was also one of the biggest believers in the power of man to change nature. He claimed that the engineered climate modification of the Sahara Sea would completely transform and ameliorate the environment in southern Tunisia and parts of Algeria. Ultimately, the French engineer had to argue for contradictory positions, however, as he also tried to appease critics who believed that the inundated chotts could adversely influence the European climate. The Sahara Sea, Lesseps thus contended, would act as a buffer between North Africa and Europe. According to this reasoning, the engineered body of water would have far-reaching local, but absolutely no global, effects. This proved to be a difficult case to defend. The logical inconsistency was simply too obvious. Some observers, like the geologist and paleontologist Auguste Pomel, generalized Lesseps's second contention, arguing that the Sahara Sea, like other large bodies of water, would probably not have any marked effect on the climate of its near and far surroundings. Thus, Egypt experienced desert conditions despite its proximity to the Red Sea and the Mediterranean. The source of the Sahara's arid conditions, Pomel argued, had little to do with water and almost all with the trade winds. Eventually, many commentators agreed that whether and how the inundated chotts would impact climatic conditions remained at least unclear, and yet the idea of modifying climates through hydraulic engineering persisted.[14]

Roudaire's project did not resolve the climate change and desertification debate. It did, however, take on a new dimension. Climate, once understood as a thing that happened only to a largely powerless humankind, had now become subject to engineering. That modifying climatic conditions was possible did not seem all that far-fetched in Western thinking in the last third of the nineteenth century. After all, if the climate had changed so drastically in the past as to cause entire continents to freeze over, and if the glaciation in Europe had, as one theory went, happened because of a flooded Sahara, steering climate by tinkering with the distribution of water could seem plausible. As in so many areas of human experience and enterprise, climatology reflected the slow shift from understanding the world as a divine and stable creation to perceiving the earth as dynamic and changeable. In the age of industrial engineering, it seemed that humanity was finally in the position to challenge or even reverse the traditional hierarchy of man and nature.[15]

This, however, did not mean that a purely mechanistic conception of the world or a radical belief in universal progress prevailed everywhere. Roudaire's notion of reestablishing *former* geographical and environmental conditions with his Sahara Sea reflected a more complex understanding of the power of technology, which could be used to mold nature in order to restore a lost golden age. Rather than being diametrically opposed to the enchanted nature of religious imagination, modern technology could take on some characteristics of the divine itself, able to actively shape the surface—and maybe even the atmosphere—of the earth. With this mindset, Roudaire was ideologically much closer to the engineers from the first half of the nineteenth century, combining Romantic and mechanistic notions of both technology and nature, than to the caricature of the modern engineer, working toward the triumph of technology *over* nature.[16]

Roudaire and many of his contemporaries did, however, argue that there was something new, something unprecedented about the capabilities of industrial technology to modify nature. Their strong belief was rooted in the new level of industrial innovation and production, which reached dizzying heights in the last third of the nineteenth century. After the first voyages of steamships, the ever expanding railroad networks, and not least the construction of the Suez Canal, it seemed only a matter of time before new technologies would overcome the biggest geographical challenges and alleviate the most pressing human concerns. In this spirit, one of the government assessors of the Sahara Sea project admitted that Roudaire's plans might seem difficult to realize but also called upon his contemporaries to not "lose sight of the fact that industry has just entered a new era in which its power rises with great speed."[17]

The projects in the Sahara also reflected a new belief in calculability, the idea that if enough data were gathered, the effects of large engineering ventures could be assessed down to the minutest details. Even most of the detractors of the Sahara Sea were not opposed to large-scale projects as such but took issue with the particular means or the dearth of quantifiable information. As in the climate change debate, data became the key to unlock a view of the future. For commentators who deemed the available facts and figures sufficient, the level of the changes that Roudaire and his contemporaries proposed did not seem disconcerting. After all, the project appealed to the widespread sense that quantifiable information would reveal the shape of future developments. To some commentators, calculation and engineering even took on a quasi-religious dimension, as the historian Dirk van Laak has argued in his study on European imperial projects: "Planning increasingly took the place of prediction, projects that of prophecies." While van Laak locates this process in the twentieth century, the same tendencies were already present in the time of Roudaire.[18]

Widespread trust in large-scale projects translated into a form of technological utopianism, which was just as prevalent in Europe as in the United States. In France, technological utopianism had a homegrown past in Saint-Simonian thought. For the disciples of Saint-Simon, technology offered to save a world that seemed to disintegrate under the contradictions between rich and poor, and unproductive and productive classes. Industrial production came to signify the central and unifying value of mankind. What better symbol for the power of industry than projects that aimed to alter the geography of the earth itself? Even better when these projects promised new settler land, where the ideals of the movement could be tested in model colonies that would develop without the constraints of traditional political and socioeconomic structures.[19] Perhaps unsurprisingly, this "religion of the industrial world" was a major source of inspiration for many of Roudaire's defenders, among them Henri Duveyrier and de Lesseps. Roudaire himself was influenced by Saint-Simonianism through his friendship with the social reformer Prosper Enfantin, who had been one of the earliest theorists of agricultural colonization in the Sahara.

Enfantin believed that colonizing the Sahara would eventually unify the Orient and Occident. Roudaire did not think quite that far but was also very confident in his own plans: "We have shown that this project does not present a single serious difficulty," Roudaire wrote about the Sahara Sea even before he had completely surveyed the chotts. Despite the mounting criticism and the increasingly unfavorable altitude measurements of the chotts, Roudaire continued to insist on the project's feasibility. When it became clear that digging the canal would require immense amounts of labor, Roudaire quickly modified his plans to include the untested and dubious suggestion that once water was introduced into the first length of the canal, its pressure would be powerful enough to cut a grade through the Algerian chotts. Similarly, once the criticism over the potential salinization of the chotts grew, Roudaire simply posited a countercurrent in the canal that would carry out the salt water to the Mediterranean.[20]

Roudaire's extreme technological optimism showed his tendency to take liberties with facts in order to fit them into his theories, a consequence of his unwavering belief in the project, which often bordered on the irrational. As each new criticism of the project revealed, the Sahara Sea was not entirely thought through, nor did it rest on a solid geographical and geological basis. The German geographer Emil Deckert pointed out the limits of the project, writing that the climatic changes that Roudaire hoped for could be achieved only if the region of the chotts were to be moved out of the "rain shade" of the mountains to the north and west that prevent moisture from moving in from the Atlantic and the Mediterranean. And even Roudaire did not think of engineering on that kind of scale![21]

Even without moving mountains, the scale and ambition of the Sahara Sea were still large enough to captivate the European, and especially the French, public. At the same time, however, there were signs that at the end of the nineteenth century the projected man-made inundation of the Sahara did not seem too surprising any longer. After the success of the Suez Canal, the Sahara Sea was facing an uphill battle to becoming the next landmark of technological progress. A commentator in *Science* wrote nonchalantly that the project of the Sahara Sea "could hardly be called a great one." The article featured a map of the entire continent of Africa, in which the region of the chotts in the far north looked relatively insignificant. Responses to the Sahara Sea came in all shapes and forms, from paeans of praise to apocalyptic warnings. That opinions were divided about a large and expensive project that remained under public scrutiny for a decade may not be surprising. Perhaps more unexpected is the scarcity of voices questioning the most basic foundations of Roudaire's project. Both favorable and unfavorable responses to the Sahara Sea displayed a common conviction that technology had become sufficiently potent to change climatic conditions and transform desert environments on a large scale. Cautious opinions that challenged this new power of mankind were rare, even among the harshest critics of the Sahara Sea.

Critical observers frequently saw the danger of large engineering projects not in human ineffectiveness but in its opposite. They feared the unintended side effects of large interventions into nature that could potentially have deleterious effects on climates and environments in Europe or even on a global scale. This frequently voiced fear illustrates the new confluence of two important strands of thought in the last third of the nineteenth century: the discussion about environmental—and specifically climatic—changes had left the academic journals and entered the public realm, where it became entwined with the debate on the potential uses of industrial technology.

The (Provisional) Demise of the Sahara Sea

Compared to the debates over the Sahara Sea's potential climatic effects, politics remained in the background of public discussions of the project. While France directly controlled Algeria, its political clout extended to Tunisia only in informal ways until 1881. That year, however, France capitalized on the growing financial difficulties of the Bey of Tunisia by invading the country, resulting in the Treaty of Ksar Said (Bardo Treaty), which made Tunisia an official French protectorate. Suddenly, the Sahara Sea had become at least a *political* possibility. Chotard's argument that the entrance to the interior sea would be in the hands of the Turks and thus useless to France—already somewhat blind to the real balance of power in the region in 1879—had now been rendered void. France was finally in the position to realize its colonial vision in Tunisia.[22]

Whether Roudaire's project was part of that official vision remained unclear. Already in 1877, the geographer Idelphonse Favé had called for more research on the feasibility of Roudaire's plans and had voted in favor of a motion to support Roudaire on further expeditions with funds from the Academy of Sciences in Paris. After Roudaire's third expedition in 1878, there was still no consensus on how much the Sahara Sea would cost. With the ongoing debates and very few clear-cut facts about the Tunisian chotts, the French government had grown less and less enthusiastic in its support for the Sahara Sea. Roudaire was further hindered in his attempts to secure official aid by the governmental instability and ministerial reshufflings between 1879 and 1883, which saw eight different governments and as many leadership changes in the Ministry of Public Works over the course of five years. Eventually, Roudaire changed his tactics, trying to further the Sahara Sea project with private funds and founding the Societé d'étude de la mer intérieure in 1882 under the directorship of Lesseps. The first step of the society was to apply for a government concession for the region of the chotts in Tunisia.[23]

With the changed political situation in Tunisia and after the ministerial reshuffles, the French government did at least attempt to reacquaint itself with the outlines of the Sahara Sea project. Prime Minister Charles de Freycinet, himself an engineer and known for his attempts as the minister of public works to nationalize French railroads, appeared to have an especially soft spot for Roudaire's plans, emphasizing the role of the Sahara Sea as a "barrier against barbarism." Freycinet called for a commission of experts to examine Roudaire's project and decide whether the government should lend further official support. With the backing of the French president Jules Grévy, the Commission supérieure was formed in May 1882. It included elected officials, delegates from various ministries, and sixteen experts, among them both known proponents and critics of the Sahara Sea. The commission was split into three subcommissions, charged respectively with determining the general feasibility of the Sahara Sea, its expected physical and climatological effects, and its expected political, military, and commercial consequences. Over the next few weeks the subcommissions held meetings and drafted a final report with recommendations to the government.[24]

The results were overwhelmingly negative. The first subcommission reaffirmed the weighty judgment of the former director of the Suez Canal Works François Philippe Voisin, who in 1881 had concluded that the Sahara Sea project was simply not feasible from a cost-benefit point of view. With the most up-to-date measurements of the desert depressions, it now seemed that the total length of canals would approximate 245 kilometers (or 152 miles), about three times as long as the Panama Canal. Moreover, the excavation would have to traverse not only the bedrock sill at Gabès, but also most of the Chott El Djerid. Voisin argued that it was practicable neither to build nor to maintain a canal of this length. And even if it were *technically* possible, building the canal would cost at least three hundred million francs, or more than four times as much as Roudaire had foreseen in even his more liberal estimates. To these charges, the first subcommission added the caveat that Roudaire's design for the canal was insufficient: as it stood, the rate of evaporation would eventually surpass the volume of water inflow, rendering the filling of the chotts physically impossible. To make matters worse, all of this work and money would be spent on a Sahara Sea that, from its original estimates of at least 16,000 square kilometers, had shrunk to about 6,000 to 8,000 square kilometers in the most recent iteration of the plans.[25]

In his final report to President Grévy in late July 1882, Freycinet stated bluntly that the French government had no grounds for encouraging the project. This assessment sounded the death knell for any further government support of the Sahara Sea. In an article published the following year, Cosson rested his case against the project by looking back at the evolution of Roudaire's plans and showing how they had become more and more complicated at every turn and more and more expensive after every expedition and every new unfavorable finding about the geological and geographical conditions of the chotts. The end of governmental support was a harsh blow to Roudaire's hopes of realizing his vision. During the debates of the commission he had very actively tried to disprove the judgments and calculations of his critics. At times, Roudaire's defense began to look absurd, and he increasingly took on the role of the blind visionary, unable to accept the less-than-convenient new findings. In his arguments, Roudaire relied more and more on simplifications of complex environmental and technological realities, falling into the trap of what James Scott has described as "abridged maps" that dissemble the possibility of simple solutions and easily moldable parameters.[26]

The number of Roudaire's followers shrank, although he could still count on the support of Lesseps, who tirelessly attempted to refute the widespread—and indeed accurate—perception that the commission's results had been deeply discouraging for the Sahara Sea project. Lesseps, who had been a dissenting member of the commission himself, stuck to his view that the plans were feasible and that the *mer intérieure* would "transform in the most marvelous way the economic, agricultural, and political conditions of Algeria." He deployed the powerful argument that before the Suez Canal was successfully completed, there had also been a plethora of naysayers declaring the impossibility of connecting the Red Sea with the Mediterranean.[27]

At this point, Lesseps had not yet squandered all of his political capital. This would eventually happen in 1892, when he was accused of bribing politicians to support a lottery trying to raise funds for the Panama Canal, which was to be Lesseps's third big venture. Even the Suez Canal itself, however, had come under increasing criticism. Despite the technological success of the canal and its opening to grand fanfare, not everything had gone to plan. For one thing, the construction had cost twice the amount originally foreseen. For another, British ships often refused to use the new and shorter shipping route, thereby drastically reducing the expected income made through tolls. Large engineering projects, while still producing great public fascination, had become a political bone of contention. Furthermore, from 1880 onward, the French government also attempted to restrict colonial expenditures, targeting hydraulic engineering projects in particular. By 1890, public works in Algeria had ground to a halt. This was in part due to the rather sorry record of French engineering projects in North Africa. Not only did dams and reservoirs fail to live up to their promises, but other hydro-engineering schemes ran into serious difficulties as well.[28]

The example of the Lake Fetzara project is telling: remarkably, the scheme represented almost the exact opposite of Roudaire's plans, if on a far smaller scale. Lake Fetzara had once been a sizeable perennial lake known for its diverse bird population, located near Annaba on the northeastern coast of Algeria (Figure 3.1). Early in the colonial period, enthusiastic French colonizers came up with the idea to use water from the lake for irrigation. The first plans for an engineering project drying up a large part of the body of water emerged

FIGURE 3.1. Lake Fetzara. *Source: L'Illustration*, August 15, 1857.

in the 1840s. Interest in the project rose in the 1870s, when one engineer foresaw the settling of French farmers on the newly reclaimed land in the lake's basin. After mining engineers theorized a link between standing water and the incidence of malaria in the region, colonial officials stressed the urgency of the project. Indeed, Lake Fetzara was drained until it was completely dry by 1880. Unfortunately for the planners, however, both the water from the lake and the basin's soil were heavily salinized. This meant that the water was unusable for irrigation, and the eucalyptus trees newly planted in the basin died within a short while.[29]

By that time, public support for big infrastructural projects in the Sahara had already taken a further, decisive blow with the disaster of Paul Flatters's expedition to the Sahara in 1881. Flatters had been on a reconnaissance mission to gather information about potential routes for the trans-Saharan railroad. In February 1881 Tuareg ambushed and killed the expedition of almost one hundred men in the vicinity of the wells of Asiou deep in the Sahara. As an immediate result of this widely reported and often-embellished episode, the French ceased their attempts to advance into the North African desert, before adopting a more aggressive form of colonialism focused on military suppression and control, spurred by a lingering desire for vengeance. Even though Roudaire underlined the military significance of the Sahara Sea, his long-term project—aiming at a form of gradual colonization less reliant on physical force—appeared out of step with the latest colonial developments. Duveyrier supported Roudaire's project precisely because it fit both with a Saint-Simonian vision of an "enlightened colonialism" and with his overwhelmingly positive views of the Tuareg of the Sahara. But in the eyes of the French public, the violent failure of the Flatters mission had proved Duveyrier—one of Roudaire's most vocal supporters—wrong.[30]

On January 21, 1885, François Roudaire died, probably of a persistent illness contracted on one of his expeditions to the region of the chotts. Lesseps remained committed to the Sahara Sea but spent more and more time planning and organizing the Panama Canal project. The society for the construction of the *mer intérieure* was taken over by Gustave Landas, who continued Roudaire's project in redirecting its main focus to the creation of man-made oases fed by groundwater from the region of the chotts. Neither Roudaire nor Landas saw their respective visions for a blooming Sahara realized, but that did not mean that schemes for reengineering the desert disappeared. When Roudaire died, the debate over large hydraulic projects was far from over. Proposals to flood parts of the Sahara and other deserts reappeared periodically, and the Western belief in the ability of modern technology to alter environments and climatic conditions—and with them the social and cultural makeup of entire regions or even continents—grew only stronger with time. In fact, even during Roudaire's lifetime, his project had not been the only one of its kind. A project in West Africa headed by the British trader and antislavery activist Donald Mackenzie had a striking resemblance to Roudaire's plans.[31]

The Long Life of the Sahara Sea

It is difficult to establish whether Roudaire and Mackenzie ever discussed their respective projects with each other. What *is* clear, however, is that Roudaire held a rather low opinion of Mackenzie's scheme, belittling it as "excessively vague." This may come as no surprise, seeing that the French engineer claimed the original idea of the Sahara Sea for himself, but even the most detached observer today would probably agree with this judgment. In contrast to Roudaire, Mackenzie had no background in engineering or geodesy. Instead, he was a merchant-adventurer who had long planned to establish trade with Africans in the south of Morocco, tapping into the long-standing rivalry between Great Britain and France over commercial interests in the region. Mackenzie had led a first expedition in 1876 to explore the region around Cape Juby, on what is today the border between Morocco and the Western Sahara in the so-called Tekna territory. After a stay in England to gather support, Mackenzie returned to the region twice, in 1878 and 1879, and managed to set up a trading station. This move provoked some hostility from both the sultan of Morocco and Spanish authorities, who saw their own respective commercial and political interests threatened by British expansion into the western Sahara.[32]

The area around Cape Juby was certainly not the most fertile or populated place, but Mackenzie was not alone in seeing a potential in the region that went beyond establishing trading outposts. In a report he wrote in 1894, the engineer Arthur Cotton commented on the possibilities to irrigate the Sahara, citing a vast fresh-water reservoir under the entire desert. Mackenzie's plan from almost twenty years earlier had the same central message but envisioned a different technique to transform the Sahara—one that looked strikingly similar to Roudaire's scheme and that would inspire a row of inland sea proposals in other parts of the British Empire.[33]

Before his first expedition in 1876, Mackenzie had already conceived of the outline for his project, which would inundate a large region in the northwestern Sahara. One reason for his trip to Cape Juby was an excursion to the western parts of the depression El Djouf (also known as El Juf), which the indigenous population had described to him as lying below sea level. El Djouf, a large expanse of sand and rock stretching deep into the Mauritanian and Malian Sahara, fascinated the British traveler and excited his inventive imagination. Over the next year, Mackenzie developed his idea of filling the depression with water from the Atlantic, ultimately aiming to make the Sahara navigable by ship almost up to Timbuktu. While Mackenzie foresaw some environmental changes in the area and even mentioned possible climatic changes, his focus was on establishing a safe trading route through the desert, which could replace the caravans and put Great Britain firmly in control of commerce in the area.

The public response in Britain to the project at El Djouf was a far cry from the excitement that Roudaire had unleashed. Mackenzie's ideas were indeed excessively vague, entirely lacking both safe geological data and an auspicious political setting in northwestern Africa. Nevertheless, the project was favorably received in some publications and often compared to the French Sahara Sea. Following the first public intimation of Mackenzie's ideas in 1875, public meetings to discuss the project took place in London, Liverpool, Bristol, and Bath. *The Times* reported the "enthusiasm" of the secretary for the colonies, Lord Carnavon, and added that, with the flooded Sahara, "Timbuctoo might become the Carthage of a tropical Mediterranean." But the newspaper also raised doubts about the plans and the lack of any concrete information on El Djouf. Despite Mackenzie's call for a survey of the depression, by 1877 its altitude had not been measured. It was neither clear how big the depression was, nor if it existed at all. Mackenzie admitted as much in his own book with the grand title *Flooding the Sahara*, which was parsimonious about the project's features while providing copious amounts of general ethnographic and economic information about North Africa. Adding to these very basic shortcomings were issues well known from the critique of the French *mer intérieure*, especially the worries about potential detrimental climatic changes in Europe. In 1878, the under-secretary of state for the colonies, Julian Pauncefote, wrote that Mackenzie's scheme was "very unlikely to be successful" and therefore did not deserve official support.[34]

Mackenzie continued to campaign for his project for a few years longer but ultimately shifted his focus to the dredging of rivers in North Africa. His new idea was to make the Sebou River navigable up to Fez, facilitating both trade and irrigation. The planned flooding of the Sahara had taken a backseat in Mackenzie's active mind. In part, this was due to the region's changed political landscape. While the 1880 Convention of Madrid had guaranteed more rights to foreigners in Morocco and thus created a more conducive climate for European trade interests, the Western Sahara, including all land to the south and east of Cape Juby, became the sole protectorate of Spain in 1885. This precluded any plans for a British-built and -controlled inland sea. The station at Cape Juby survived a little longer but remained under the constant threat of Moroccan attacks and political intrigue—or at least that was how Mackenzie portrayed the situation. Eventually, the sultan of Morocco purchased back the land and Mackenzie refocused his energies on antislavery activities, first in North Africa and then increasingly in East Africa, where British influence was growing rather than receding.[35]

The Sahara seas of Roudaire and Mackenzie had nevertheless already made a lasting impression that reached beyond the borders of Europe. The American press discussed the projects with some interest, not least because of similar ideas for improving the arid plains and deserts in the American West through large-scale, government-supported irrigation measures. There was even a plan— possibly inspired by Roudaire's ideas—to use water from the Gulf of California to flood the Colorado Desert. By the second half of the 1880s, however, the Sahara Sea projects started to disappear from the news and slipped into half oblivion. A rare 1895 article on the Sahara Sea by Paul Staudinger already described the project as one of the distant past. And yet, Roudaire's endeavor had done much to put a part of Algeria that had been little known to Europeans into the limelight and had added to insights in geodesy, geology, botany, anthropology, archaeology, and mineralogy in the region. This was the real accomplishment of the Sahara Sea, as Chotard—critical of the project's feasibility—had already foreseen in 1880. Moreover, oblivion was never complete. The Sahara Sea resurfaced in the news sporadically, as Roudaire and Mackenzie had managed to inspire a new generation of engineers and entrepreneurs to further develop the ideas of large hydro-engineering projects in desert regions.[36]

In the 1890s, two French hydro-engineering projects continued Roudaire's legacy, one suggesting the formation of man-made oases at Oued Righ (also transliterated as Oued Rirh) for date farming in southern Algeria and the other proposing a canal from Cape Juby to Timbuktu. Despite the lack of acknowledgment, the latter plan was clearly based on Mackenzie's ideas. An argument over authorship never materialized, however, as the French government was very quick to turn down the project. After the turn of the century, the French engineer Etchegoyen once again considered the flooding of part of the Sahara.

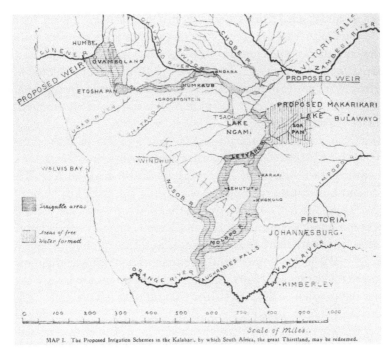

MAP 1.   The Proposed Irrigation Schemes in the Kalahari, by which South Africa, the great Thirstland, may be redeemed.

FIGURE 3.2. Ernest Schwarz's Kalahari project. *Source*: Ernest H. L. Schwarz, *The Kalahari or Thirstland Redemption* (Cape Town: T. M. Miller, 1920).

He was convinced not only that the chotts and El Djouf were former seabeds but that a full quarter of the Sahara's surface lay below sea level. Like Roudaire, Etchegoyen hoped that flooding the Sahara would ameliorate the climate, but newspapers reporting on the project repeated the earlier caveat of a potential climatic catastrophe in Europe and added the perceptive warning that redistributing copious amounts of water from the sea to the desert might even induce the earth's axis to shift.[37]

Roudaire's plans found a new fully fledged iteration in 1920—this time thousands of miles south of the chotts. There, the South African engineer Ernest Schwarz published his plans for "Thirstland Redemption," a large hydro-engineering project in and around the Kalahari. Schwarz intended to flood the Etosha Pan in modern-day Namibia, enlarge the endorheic Lake Ngami, and create a large body of water in the Sua Pan in modern-day Botswana (Figure 3.2). The scheme, Schwarz argued, would transform the dry lands into cultivable acreage and counteract the increasing desiccation of South Africa, a contention the engineer backed up with historical information on failing harvests and the disappearance of large animals from the area, as well as rainfall data. In a way, Schwarz had translated the Sahara Sea projects to suit South African conditions, using a very similar rhetoric about environmental decline and emphasizing the project's value to the white settlers. Even Schwarz's choice of location, between a British colony and the newly acquired territory of Southwest Africa, was reminiscent of the region of the chotts between the old French colony of Algeria and the newly acquired protectorate of Tunisia.[38]

The Sahara itself also remained popular among engineers. In the 1920s the American businessman Dwight Braman came up with a plan to irrigate a large part of the desert. And in the early 1930s the British engineer John Ball expanded on earlier plans to flood two wadis in Egypt. Ball's first choice for a large-scale inundation, however, was neither of those wadis but the Qattara Depression, about 130 miles west of Cairo. Despite his frank admission about the shortage of data, Ball proposed engineering interventions on a large scale. He suggested digging a tunnel to the Mediterranean, which would allow water to flow to the depression. Ball's primary objective was not to induce climate change but to produce hydroelectricity with turbines in the tunnel. Ball also referred to a plan of diverting some water from the Nile through its alleged former western riverbed to irrigate the Sahara. In 1936, then, the Danakil Depression in the Afar Triangle of Ethiopia became the target of a short-lived inundation plan with the aim of creating a harbor for the export of oil from the Aussa region.[39]

Roudaire's project continued to feature as a model and inspiration. In 1928, the American popular science author Edwin Slosson referred to it directly in a bitingly sarcastic article on the recent enthusiasm about the alleged discovery of Atlantis in "the salt swamps of Tunis." This was a stab at the German Paul Borchardt, who claimed to have located the lost city in the region of the chotts and argued that the North African climate had deteriorated in historical times—one of the reasons, he argued, for the disappearance of Atlantis and its high culture. The climate change debate had certainly come a long way. And the suggestion that Atlantis had been in the region of the chotts added new credence to projects that anticipated the restitution of an alleged prior fertility.[40]

Reporting on this search for Atlantis, Slosson had nothing but sarcasm for his contemporaries who tried to locate the city in the chotts, but he still cultivated the idea that the area had once been marked by a much more fertile climate: "Why the country once so flourishing should now be so desolate is a disputed question," he wrote. "Some lay it to the increasing aridity of the climate. Others say it is due to volcanic disturbance of level. Others ascribe it to the 'blight of Islam.'" With that, Slosson had quite neatly summed up the different theories on climate change that had been formulated in the nineteenth-century debate: a great universal process, a local catastrophic process, or the effect of human actions. While the academic debate on climate change had entered a quieter period in the 1920s, it lived on both in the debates about the lost city of Atlantis and in great projects of environmental transformation. These projects also inspired popular culture, such as George Griffith's novel *Great Weather Syndicate* about a scheme to influence global weather. The book closes with a scene at Tan-ez-Ruft, "one of the most terrible of Saharan regions" now transformed into fertile land, "blossoming as the rose." This kind of triumph, possible in fiction, remained out of reach for Roudaire and his successors.[41]

At the end of *The Invasion of the Sea,* Jules Verne seems to not leave us with many options to interpret his story. The central moral becomes all too clear after an earthquake rocks North Africa, kills its inhabitants, and finishes what humans had only dreamed of. Nature is stronger than man's vision and will. And yet in Verne's story nature ultimately helped to complete what humans had already started to design. In fact, *The Invasion of the Sea* suggests that nature and engineering actually cooperate to adjust the environment to fit human ends. This is also what Roudaire hinted at when he emphasized that the chotts had been filled with water in the past and that his project sought to re-create a former, more perfect condition of nature.

Despite its status as an unrealized project, the Sahara Sea is a noteworthy chapter in the European chronicles of colonial encounters with non-European environments. Roudaire's plans, along with those of his contemporaries and successors, show not only that new scientific ideas about the environment were closely bound up in colonial discourses and endeavors, but also how European colonialism came to be characterized by visions of transforming environments and climates and by a new level of cornucopian ideology. The age of engineers had only just begun in the late nineteenth century. Roudaire's ultimate failure to execute his plans did not deter his colleagues and successors but seemed only to spur them on. The question was not so much *if* a project like the Sahara Sea would work but rather *when* the right technologies would be in place to realize it. The fast pace of technological innovation and organization made a breakthrough seem imminent at any moment. And the threat of detrimental environmental and climatic change ensured that desert transformation remained on the agenda.

If Roudaire had been the archetype of the prophetic Faustian engineer in the nineteenth century, the German architect Herman Sörgel filled that same role in the early twentieth. His Atlantropa project took Roudaire's plans as inspiration and a point of departure but elevated them to an altogether new extreme of continental geoengineering, which proposed the complete transformation of the landscapes and climates of Africa.

4

A New Climate for a New Continent

HERMAN SÖRGEL'S ATLANTROPA

IN 1997, THE GEOPHYSICAL MAGAZINE *EOS* published a short article by R. G. Johnson, proposing a bizarre-sounding geoengineering project. Johnson suggested the construction of a dam at Gibraltar, inserting a partial barrier between the Mediterranean and the Atlantic Ocean. According to the article, the dam would alter oceanic currents, ultimately preventing the allegedly impending onset of a modern ice age. Johnson did not explain fully how glaciation dynamics meshed with global warming and was, in general, unconcerned about potential problems in his rather schematic climate models. Sure enough, two of Johnson's colleagues promptly criticized the article, citing the lack of both scientific rigor and quantitative data to back up his daring hypotheses. And this is where the debate stopped. Johnson's article turned out to be no more than a curious episode in the publishing history of *EOS*. Possibly unbeknownst to Johnson himself, however, the proposed dam at Gibraltar was actually not a new idea. In his article, Johnson had re-created not only the main physical feature of an engineering project from the 1920s but also the attendant rhetoric and the technological enthusiasm of its creator, the German architect Herman Sörgel.[1]

Sörgel's project, first called Panropa and later Atlantropa, also centered on a giant dam at Gibraltar (Figure 4.1). In contrast to Johnson's plans, however, Sörgel proposed a complete barrier between the Mediterranean and the Atlantic Ocean, linking Europe and Africa. "Atlantropa" was Sörgel's own neologism, combining "Atlantic Ocean" with "Europe" and evoking the lost city of Atlantis. At least on paper, however, Atlantropa actually surpassed the mythical city: upon construction, it would expose new land around the Mediterranean basin, change climatic conditions as far afield as Northern Europe, and produce vast amounts of hydropower to fuel the irrigation and complete environmental transformation of North Africa. Sörgel's ideas represented an idiosyncratic mix of colonialist conceptions, racist ideology, cultural pessimism, anxieties about imminent fossil fuel shortages and global

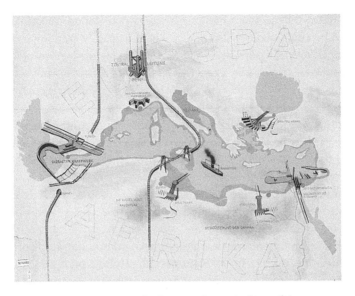

FIGURE 4.1. Map of Atlantropa. *Source*: Archives of the
Deutsches Museum in Munich.

environmental decline, and a cornucopian belief in technology and postnational cooperation. From today's vantage point, Atlantropa may seem merely
like a bizarre sideshow to the momentous events of the interwar period. Despite its unique scale and unrivalled hubris, however, the project was no anomaly in its time and place. Atlantropa became a topic of public debate in Germany and beyond, fitting well into the politically and culturally volatile years
of interwar Europe: like so many other projects of the time—from new political ideologies to attempts at reenvisioning modern ways of living—Atlantropa's
creator endeavored to find an escape from a seemingly all-encompassing crisis
that had taken hold of the continent.[2]

Atlantropa reflected similarly far-reaching goals of other interwar projects.
With his scheme, Sörgel envisaged a radical political and social reorganization
of Europe. Moreover, he participated in the wider discussion about the role of
technology in human interactions with nature, expressing his concern about
what he perceived to be the shortcomings of conventional politics in exploiting new technological possibilities. The prevailing political structures, Sörgel
argued, were outdated and unable to use the available tools and expertise to
mold nature according to human—more specifically European—needs. The
architect thus conceived Atlantropa not only as a giant engineering project
that would provide energy for centuries to come but also as a revolutionary
endeavor that would radically transform environments and overcome the

nation-state as the primary building block of political organization. Sörgel believed he could achieve this ambitious end goal by creating new living space for a European population that was stifled on an overcrowded and divided continent. Africa, the "expectant vacuum in front of Europe's doors" in his words, was the obvious target. Before any large-scale settlement would become possible, however, Europeans would have to make the continent suitable for colonization. According to Sörgel, this would require above all transforming the hostile climatic conditions in large parts of Africa and stalling the progressing *Versteppung*, or desertification, that was threatening human existence.[3]

Atlantropa thus embodies a prime example of how theories and terms that had been developed in the academic debate on climate change were politicized in the early twentieth century. It also shows that both the common fears of large-scale environmental decline and the even more widespread enthusiasm for engineering solutions to "fix" nature endured. Desiccation and desertification remained powerful concepts and became increasingly attached not only to environmental but also to social and cultural theories and anxieties. Herman Sörgel, whether deliberately or not, used the terms of the climate change debate to advertise his grand vision of both an engineered new continent and an engineered new society, ready to stall and then reverse the advancing environmental and cultural decline of the West.

Postnational Hydropower

Herman Sörgel was born in 1885 in Regensburg. His father, Johann Sörgel (1848–1910), had been the board's director of the Bavarian Building Commission, which had initiated the construction of the Walchensee Power Plant in southern Bavaria, designed by the politician and engineer Oskar von Miller. Around the turn of the century, the Sörgel family moved to Munich, and Herman enrolled at the local Technical University after graduating from secondary school. He received a graduate degree in engineering, but the faculty declined to grant Sörgel a doctorate on the basis of his study on architectural aesthetics—a topic that his instructors deemed too far removed from the engineering curriculum of the institution. The dissertation did, however, receive a warm reception from the famous architect Fritz Schumacher at Dresden University, who attempted, but ultimately failed, to procure an honorary degree for Sörgel.[4]

Giving up on his pursuit of higher academic honors, Sörgel followed in his father's footsteps by taking up a position in the Bavarian Building Commission. Among other projects, he designed the building for the hydropower station in Aufkirchen on the Isar River. During and immediately after the First

FIGURE 4.2. Herman Sörgel posing for a photograph in front of
Atlantropa maps. *Source*: Archives of the Deutsches Museum in Munich.

World War, Sörgel also started to publish his first books, among them a well-
respected theoretical treatise on architecture. For a while he was also the
editor-in-chief of the architectural journal *Baukunst*, a position that put Sörgel
in contact with leading architects who would later contribute to his plans to
geoengineer the Mediterranean.[5]

If we can trust Sörgel's own account, the idea for Atlantropa came to him
in a sudden flash of inspiration in 1927. Sörgel had just finished reading a treatise on the Mediterranean by Otto Jessen, a German geographer who was also
taking part in the then fashionable hunt for the lost cities of antiquity. Building
on nineteenth-century research, Jessen characterized the Mediterranean as an
"evaporation sea." Its water level, he argued, was kept up through the constant
inflow from the Atlantic Ocean through the Strait of Gibraltar. This mundane
hydrological detail would provide the basis for Atlantropa's most prominent
feature, which was as simple as it was grandiose: a giant impermeable barrier
connecting Morocco and Spain at Gibraltar. This oversized dam would cut the
water supply from the Atlantic Ocean, while smaller, yet still sizeable, dams at
Gallipoli and the Nile estuary would isolate the Mediterranean from other
large bodies of water.[6]

Through slow evaporation, Sörgel assumed, the water level of the Mediterranean would drop, eventually exposing large swaths of former seabed around the coastline. In a next step, Sörgel envisioned giant hydropower generators on the dam that could use the buildup of potential energy between the now different water levels of the Atlantic and the Mediterranean. This basic design—the "Mediterranean as the power plant of Europe"—would remain the most iconic feature of Atlantropa, although the project later included the entire continent of Africa. The main thrust of the plans that Sörgel laid out in detail in his first book on Atlantropa in 1932 would remain largely unchanged over the next few decades. Throughout the political upheavals from the 1930s to the 1950s in Europe, the architect stayed true to his ideas and worked tirelessly to promote the project, to which he would devote the rest of his life.[7]

Atlantropa was never exclusively an architectural or engineering project. From the beginning, Sörgel conceived of his scheme as a revolutionary force that would reorganize and restructure Europe socially and politically as well as physically. Atlantropa was to be "a new life form of Europe," uniting its quarreling constituent nations and redirecting their energies from internecine warfare to a great collaborative project. In 1925, Sörgel had visited the Pan American Union building in Washington, D.C., and was struck by this tangible manifestation of international cooperation (Figure 4.3). Although the union, the predecessor of the modern-day Organization of American States (OAS), was far removed from Sörgel's ideal of an all-powerful centralized association, he still viewed it as a model for his political plans in Europe.[8]

Even closer to the essence of Atlantropa was H. G. Wells's voluminous *Outline of History*, which became another source of inspiration for Sörgel. Under the fresh impression of wartime destruction, Wells had criticized the feeble attempts at international cooperation through the League of Nations and called for a "world league of men." Before this end stage of postnational history could become a reality, Wells foresaw associations on a smaller level as steps toward the ultimate goal. He mentioned an improved Pan American Union and spent a few lines describing his idea of a United States of the Old World. Sörgel's later rhetoric about the "three A's"—or Asia, America, and Atlantropa— was similar to Wells's concept of a world divided into large supranational blocs. Pondering the future world state, Wells imagined a singular economy built upon scientific principles that would benefit the entirety of humankind. The founder of the Pan-Europa movement, the Austrian-Japanese politician Richard Coudenhove-Kalergi, mirrored these thoughts. He called for a cosmopolitan and technological reorganization of politics and understood the First World War as a civil war between Europeans, who had forgotten their common destiny and had misused the great technological advances of the industrial age.[9]

FIGURE 4.3. The Pan American Union in Washington, D.C. (1943).
Source: Public domain, but originally from the Library of Congress.

Sörgel openly praised some of Coudenhove-Kalergi's ideas but found them insufficiently ambitious. He nevertheless adopted a strikingly similar attitude toward technology and the use of natural forces as he elaborated his plans. Some of the first calculations for Atlantropa estimated the volume of water inflow from the Atlantic to the Mediterranean at the Strait of Gibraltar to be about twelve times the amount of water that rushes down the Niagara Falls. For Sörgel, it was incomprehensible that people had not yet made use of the potential energy stored in the Mediterranean, particularly in light of his calculation that coal supplies would be depleted in a few hundred years. Later, Sörgel added to this estimate that global energy consumption was doubling every twenty years, further illustrating the unsustainable and "destructive exploitation" of coal reserves. His anxieties about the exhaustion of fossil fuel supplies were, in fact, one driving factor behind his plans for Atlantropa. Sörgel warned his readers that coal deposits would be depleted in a couple of centuries and that the store of fossil oil would last only for another two decades. Sörgel tapped into long-standing worries about fossil fuel shortages that had been around in Europe since at least the late eighteenth century, had found notable expressions in the dire predictions of William Stanley Jevons in the nineteenth century, and picked up an even more urgent tone in the early twentieth century.[10]

After the energy shortages of the First World War, the search for alternatives to fossil fuel grew even more urgent—with ideas for harnessing the power of rivers, oceans, and the sun cropping up in various places. In countries with a fair share of mountainous terrain, hydroelectricity continued to be the focus of development. "White coal" had become a viable alternative to fossil fuels around the beginning of the twentieth century, when efficient power lines made it possible to transport electricity cheaply from the sites of production to the places of consumption. In fact, the mastermind behind the Walchensee Power Plant, Oskar von Miller, had played a major role in this development by conducting the first successful public experiments proving the practicability of long-distance transmission of electricity in the 1890s. This finding inaugurated a long period of enthusiasm for hydropower, which various people and interest groups advertised as the solution to both economic and social problems around the turn of the century.[11]

The zeal showed few signs of subsiding in the 1920s. Italy, the most impressive case study in this respect, had already led Europe in hydropower usage before the war but embarked on an unprecedented expansion of its resources in the north under the new fascist regime in the 1920s: between 1926 and 1935, hydroelectric production almost doubled, making up for more than 95 percent of electricity production in the country! Other European countries developed hydropower at a far slower pace, but the fervor surrounding this "new" form of energy was widespread even there. Far removed from the wooden watermills of the preindustrial era, hydropower in the twentieth century presented itself as a technology at the forefront of progress: the production sites, with impressive dams and oversized penstocks and turbines, were feats of modern engineering. Hydropower was advertised as clean, scientific, limitless, and—because of the small staff required to run power stations—"practically free from labour troubles," as one participant of the first World Power Conference in London put it.[12]

For countries such as Germany and France that had suffered from fuel shortages during the First World War, hydropower became part of a strategy for greater energy independence in the years thereafter. Although hydroelectricity was by then no longer a new technology, enthusiasm for it remained strong. And waterpower advocates found an eager audience among the German population, who appreciated any kind of optimistic economic projection during the hardships of the postwar period. Even clearly inflated assertions, like that of the engineer Heinrich Voegtle, who claimed that 70 percent of Germany's energy demand could be satisfied with hydropower, were more than welcome in the Weimar Republic.[13]

More sober observers cast a less favorable light on hydropower's potential. The engineer Arthur Lichtenauer calculated in 1926 that a complete change from coal to hydropower was out of the question. As he argued in his doctoral dissertation, every country in Central Europe had a higher total energy demand than its respective potential hydropower supply. Lichtenauer did not fully disclose how he had calculated the potentials of each country, nor how he had accounted for possible increases in potential through constructing dams and developing more efficient penstocks and turbines. The message, however, was clear enough: hydropower alone could not make up for future energy shortages caused by the ongoing depletion of coal reserves. Lichtenauer put his faith in making the most of other renewable energy sources through tidal generating plants and wind power stations. Europe simply did not have sufficient stores of water with high potential energy that could be exploited for hydroelectric power generation.[14]

Herman Sörgel probably would have agreed with Lichtenauer's note of caution: in the confines of an unaltered geography, Europe did not have sufficient energy resources to compete with the rest of the world or even to keep up its level of industrial production and human reproduction. Atlantropa, however, could solve this dilemma, Sörgel argued, because it would *create*, rather than simply *use*, potential energy. The first calculations projected a capacity of 160 million horsepower (about 120 gigawatts) at the Gibraltar dam alone, equaling the output of about one thousand Walchensee power plants. But for Sörgel, Atlantropa was more than a welcome source of energy for centuries to come; it was also an assurance against further intra-European conflict. An integrated power grid would be a better guarantee of peace than any written treaties, Sörgel argued, as any damage to the lines would be a detriment to every nation. Even more than that: energy came to signify a promise of survival for a continent that Sörgel saw as being on the brink of demise. In a theme that he would come back to throughout his life, he insisted on the link between energy, environment, and culture: if Europe did not want to be outflanked by other continents, Sörgel argued, it had to develop the Mediterranean as a power source. Otherwise, "the cultural center of Europe would vanish. Europe would become desolate [*veröden*], paralyzed; at best it would still retain a fossilized culture, like Egypt or India today."[15]

This fear of European decline was at the core of Atlantropa. It was an all-encompassing decline that Sörgel feared: the anticipated energy shortage was a symptom of a comprehensive malaise that had taken hold of the continent.

Sörgel's mixture of images and terms to describe his project—borrowed both from the natural sciences and from reactionary German philosophy—was no accident. The "desolation" of Europe was always both a material and a spiritual threat that manifested as environmental and cultural decay. The continent, Sörgel believed, was threatened by desertification of the soil and the mind, and the only way out of this predicament was for its people to expand beyond its geographic borders, ensuring both new energy sources and an enlarged territory that would overcome the stifling overcrowding of Europe. Atlantropa offered a bridge to a promised land across the Mediterranean. This land was the Sahara, itself the victim of desertification on a vast scale, as Sörgel repeated again and again in his writings. But it offered the space that the overpopulated Europe needed to survive, and it also offered vast testing grounds for modern engineers to prove technology's ability to alter environments, climates, and societies.

In the Footsteps of Roudaire

When Sörgel first introduced his ideas to a wider audience, the Gibraltar dam quite naturally became the focus of attention. The unprecedented scale of this projected structure, and the almost biblical audacity of separating two seas, piqued the interest of the public. In fact, the Gibraltar dam—like other large hydro-engineering projects—would provide the material for a number of more or less tawdry science-fiction novels in the 1930s, most of their plots including attempted sabotage acts during or after the construction of the sea barrier. The dam was certainly a key element of Atlantropa, but Sörgel himself emphasized a different part of the project. The dam and its hydropower generators were only a means to the end of reclaiming territory for an all-European colonial settlement, which in turn would dissolve nationalist tensions on the continent. The newly exposed land around the Mediterranean littoral would be a small part of this plan, while the heart of Atlantropa lay in North Africa, which would provide vast amounts of space for European expansion. Sörgel's ultimate aim, as he explained bluntly in his first book on the project, was "the domination of the black continent by Europe." This would be possible, however, only if the environmental and climatic conditions of North Africa could be changed to suit the demands of white Europeans.[16]

Sörgel explained to his audience that the economic potential of Africa consisted not only in what could be achieved by improving soil conditions but also in what could be accomplished through "geographical alterations [and] climatic improvements." On a less material level, Sörgel also mentioned a "natural law" that answered to the human "*horror vacui*," the fear of emptiness: "Sooner or later, a vacuum like the Sahara has to be filled. Regardless of whether with water or humans." His plans for Atlantropa, in fact, called for both of these alternatives. With the energy produced at the hydropower stations on the dams around the Mediterranean, Sörgel intended to pump water from the sea into the Sahara. The water would create large navigable lakes and transform the environment and climate through new levels of precipitation. Free of the heat and aridity of the Sahara, the transformed landscape would then be "filled" with the surplus population of Europe.[17]

This plan not only exemplified colonial rhetoric about empty African spaces but also was strikingly reminiscent of François Roudaire's project in southern Algeria. This was not a coincidence. Sörgel knew of the French engineer's plans to flood part of the Sahara and cited them as a model. He adapted Roudaire's optimism that man-made inland seas would ameliorate the desert climate and allow for settlement and agriculture. Sörgel also referred to the salt lake el Djerid—Roudaire's choice for his intended *mer intérieure*—as one of the most promising floodable areas within the Sahara. Sörgel estimated that the feeder canal from the Mediterranean would require five years and thirty million dollars to build, carefully omitting the mounting problems that his French predecessor had encountered.[18]

Sörgel also mentioned other similar projects in support of his ideas, among them Donald Mackenzie's Sahara Sea, the Qattara Depression plans by John Ball, and an obscure scheme of the Boston banker Dwight Braman. Braman had first caused a stir in 1897 when he proposed to President McKinley that Cuba be converted into a privately held, mixed Spanish-American corporation. In the 1920s he was the director of the "Allied Patriotic Societies," a jingoistic anti-immigrant and anticommunist group. Maybe as part of his plan to divert immigrants from coming to the United States, Braman had incorporated a company in France whose main aim was to flood 100,000 square miles of the Sahara, creating fertile land around the new coastline for up to 4.5 million families. Braman had claimed in 1928 that both the French and Italian governments were interested in his project. Whether this was true or not, his premature death in France the following year nipped the implementation of his plans in the bud.[19]

In fact, none of the Sahara projects Sörgel cited in his work had actually been realized, but he remained convinced that this was only due to a lack of will. Sörgel never doubted that his plans to cultivate the Sahara were feasible, and he believed that the time had come to realize them. In support, Sörgel cited the Austro-Hungarian botanist and philosopher Raoul Heinrich Francé, who had described the "re-planting of forests" in the Sahara as the most promising cultural task of the twentieth century. As Francé's phrasing reveals, he imagined this process as a restoration of the North African environment to its allegedly original state: the Sahara was to be "*re*-planted" to convert it back to the fertile landscape it once had been. In his travel writings, Francé appeared shocked by the extent of deforestation in the Mediterranean region. Assigning the blame to proto-capitalist modes of production, he stressed the anthropogenic cause of this loss of the "natural equilibrium," which he traced back to the Carthaginians and their wood-intensive attempts to build a great merchant navy. Sörgel himself wholeheartedly subscribed to the notion of a sharp environmental decline in the region and propagated the image of a formerly productive North Africa.[20]

Sörgel's second big publication on Atlantropa was even more outspoken. Not only had the desert once been fertile, he argued, but desiccation was still under way at the alarming rate of one kilometer per year. While Sörgel remained quiet on the cause of the growing Sahara, he was interested in the signs and effects of that process: "Eradication of the forests; heating up of the soil; drying up of the air; desertification [*Verwüstung*]." Echoing French colonial administrators, Sörgel referred to North Africa as the "granary of Rome," which he aimed to resurrect to ensure the survival of the European successors of the once mighty empire. The views of the late nineteenth century had evidently carried over to the twentieth. They had, in fact, become so engrained that their dubious evidentiary basis was rarely challenged.[21]

Scholarly debates about the existence and causes of large-scale climate change that had been so prominent in geological journals around the turn of the century had run out of steam when Sörgel was conceiving Atlantropa. Concerns about desiccation, however, had survived the war and may even have been heightened by the accompanying food and fuel shortages in Europe and the stirrings of anticolonial movements in Africa and Asia. In the 1920s, Ellsworth Huntington and Charles Brooks published popular studies covering large-scale desiccation and climatic decline. And in the 1930s, the widely covered Dust Bowl in the United States brought desiccation to the forefront of the debate once again. The Sahara, the proving grounds for explorers and climatologists in the nineteenth century, also remained a topic of discussion. As the largest desert close to Europe, it provided powerful images for the construction of a cautionary tale on the extremes of desiccation and erosion.[22]

The British colonial forester Edward Stebbing was one of the most important voices in the renewed desertification debate. As the former head of the Indian Forest Department, Stebbing had published a three-volume compendium on the past, present, and future of forests on the subcontinent in the early 1920s. During his career as a colonial administrator and then as professor of forestry at Edinburgh University, Stebbing traveled across other parts of the British Empire and wrote on what he perceived to be the steady destruction of forests and the growth of deserts. In "The Encroaching Sahara," one of his most discussed lectures, he argued for anthropogenic causes like overgrazing and burning and ended with the pointed question: "How long before the desert supervenes?" Following his predecessors in the nineteenth century,

Stebbing failed to see or mention the interconnection between colonial occupation, the displacement of indigenous populations to ever more marginal
lands, and the increased rate of deforestation.[23]

Like Sörgel, Stebbing believed that halting the advancing desert was well
within the reach of available means. Citing measures in the United States as
examples, Stebbing proposed a protective and protected forest belt that would
arrest the "silent invasion" of the Sahara. Two years later, he repeated both his
warning about desertification and his proposed solution, this time focusing
even more closely on the anthropogenic reasons for the Sahara's encroachment and sounding an even louder alarm about the high speed of the process.
In the same article, Stebbing referred back to the discussions about climate
change that had begun in the late nineteenth century, describing them as still
unresolved and in need of further investigation. Apparently, he was willing to
consider causes beyond human action for desertification, which in any case
still had not been conclusively defined. Stebbing himself acknowledged as
much in his writings, revealing the ongoing confusion about both the terminology and the causes of desertification in the 1930s.[24]

Stebbing was not alone in calling out the threat of the Sahara. Whether
anthropogenic or natural, desertification remained an enduring topic of discussion. Reminiscing about work on the large British *African Survey*, the Cambridge biologist E. Barton Worthington remembered the contributors' focus
on "the southern creep of arid conditions" threatening agriculture in West
Africa. In Germany, one newspaper reported in 1935 on the "alarming news"
about the "inexorable advance of the desert to Europe," adding that the reclamation of the Sahara as a place for settlement was one of the most pressing
tasks of the century.[25] Maybe unsurprisingly, this newspaper article made it
into the collection of clippings on topics related to Atlantropa that Sörgel meticulously maintained through most of his life. Desertification—not bound by
political borders and possibly threatening the entire European continent—fit
well with Sörgel's justification for an overarching project of a new political and
cultural order. If the threat transcended nations, then the solution would have
to be postnational as well. To highlight the significance of the challenge, Sörgel
had been trying to collect data on the meteorological and climatological conditions in North Africa from the very early stages of his project. While he did
not seem to have any substantial doubts about the feasibility or eventual success of his project, he knew that he had to convince others. In 1931, he attempted to solicit information on hydrological conditions in the southern part
of the Mediterranean basin from the University of Pavia in Italy. In the end,
the Swiss engineer and world traveler Bruno Siegwart supplied Sörgel with
some of the desired information. Siegwart eventually also became one of the
most enthusiastic supporters of Atlantropa and the coauthor of the second

stage of the project, which focused almost entirely on the transformation of Africa.[26]

Beyond the Granary of Rome

Sörgel first became acquainted with Siegwart through an article on Atlantropa in a Munich newspaper in 1928. The Swiss engineer had read the short description of the project and was intrigued yet unconvinced. He wrote a letter addressed to the newspaper warning Sörgel and the public that the dam at Gibraltar might lead to problems and even catastrophes on an unforeseeable scale. Siegwart expected a considerable rise of ocean levels through the drop in the water level in the Mediterranean basin: if the level behind the dam was lowered by as much as several hundred meters, as the first plans anticipated, the displaced volume of water would find its way into the world's oceans and cause sea levels to rise up to one and a half meters—a serious issue Sörgel had neglected to address up to that point. In a second letter—now addressed directly to Sörgel—Siegwart repeated the fear of rising oceans, but added another, even more serious warning: the redistribution of water, he argued, could cause the axis of the earth to shift.[27]

This contention was difficult to prove but not to be dismissed lightly: after all, a shift of the earth's axis could lead to drastic, even catastrophic climatic changes all around the globe. To add insult to injury, Siegwart was skeptical about Sörgel's idea of transforming the Sahara. He feared that the dream of an irrigated and fertile desert might ultimately remain an illusion: even modern technology would be unable to overcome the geological realities of rocky and sandy surfaces lacking any significant layer of topsoil. And yet Siegwart included plans and drawings of harbors, canals, and power stations for the reengineered Mediterranean in the very same letter. Despite his grave doubts about the project, his curiosity about Atlantropa was winning out.[28]

Sörgel addressed Siegwart's fears rather feebly. In a letter to the editor of a newspaper in Cologne, Sörgel turned to an esoteric notion of miraculous natural processes that would restore the balance of the earth. The sea level rise was only conjecture, he argued, as the global water cycle did not work in purely mathematical ways. With a very similar argument, Sörgel attempted to preempt the critique in his first major publication on the project: over the span of the two hundred years that lowering the Mediterranean was projected to take, the volume of water represented but a negligible amount in the global water cycle and would, in fact, be subsumed by "imponderable processes of nature." The shift of the earth's axis, Sörgel contended further, would be only minimal. He conceded that the redistribution of water might lead to a higher incidence of earthquakes—this was something Sörgel could hardly ignore, as

FIGURE 4.4. Planned dam between the basins of the Congo and the Chad Sea.
Source: Herman Sörgel and Bruno Siegwart, "Erschliessung Afrikas durch Binnenmeere:
Saharabewässerung durch Mittelmeersenkung," *Beilage zum Baumeister* 3 (1935): 37.

increased seismic activity around large bodies of water had been documented
ever since the first big reservoirs were created in the nineteenth century. Sörgel
tried to alleviate possible worries by alleging that the slow lowering of the
water level in the Mediterranean could always be halted if earthquakes became
dangerous to the structural stability of the dam. This would have been cold
comfort for future Atlantropans living in a trembling Mediterranean basin, but
it seemed to win Siegwart over, as he began to work on an expansion of the
project beginning in 1932–33. The focus of this next part of the project was on
central Africa, where the two collaborators now started to envision an engineered lake of gigantic proportions.[29]

Their first exchange of ideas about these new plans took place as early as
1932, when Siegwart came up with the proposal to channel water from the East
African lakes into the Sahara for irrigation. A year later, the conversation had
shifted westward, and Sörgel and Siegwart discussed the feasibility of damming the Congo River to create a lake in central Africa. The idea was to enlarge
the Stanley Pool to twenty-four times its original size, creating a body of water
with a surface area of about 12,000 square kilometers, a volume of 55 cubic
kilometers, and the potential capacity of 22.5 to 45 gigawatts of hydropower
(Figure 4.4). In contrast to the Gibraltar dam, the "weir" at the Congo would
be "a dwarf" and require far less effort to construct, Siegwart wrote. He also
identified a natural basin for this engineered lake, with mountains or higher

terrain surrounding the projected site. In an inverse argument to Sörgel's claim that the Mediterranean had once been dry and inhabited, Siegwart now asserted that the Congo basin had once been a sea. This was not the only respect in which the plan resembled Roudaire's 1870s project. Siegwart and Sörgel also thought about channeling water from the newly created Congo Lake to an enlarged Lake Chad and, from there, to the Chott el Djerid, creating a connection to the Mediterranean. Fifty years after Roudaire's death, this step would have finally realized the French engineer's dream, re-creating the alleged Lake Tritonis in the Algerian and Tunisian Sahara.[30]

Sörgel and Siegwart published these plans for the first time in 1935. The article included not only the plans for the Congo Lake but also a map showing the enlarged Lake Chad and a third engineered body of water in southern Africa, which would be created by damming the Zambezi (Figure 4.5). Atlantropa had now grown from an already large project centered on the Mediterranean Sea to geoengineering on an unprecedented scale. To support their gigantic project, Sörgel and Siegwart cited data from forty Belgian meteorological stations in the Congo and topographic maps of the Oceanographic Museum in Monaco. They estimated that filling the basin would take "about 133 years," an estimate that obscured the ongoing difficulties that the two planners encountered in collecting detailed information on central Africa: in private correspondence, Siegwart mentioned the lack of data on topography and total evaporation loss, the incidence and volume of exceptional floods, and meteorological information on winds. The shortage of reliable information did not stop Sörgel and Siegwart from describing the expected effects of this reengineering of the African continent, foreseeing the cultivation of the northern regions of the Sahara with the collected water from the Congo River. This step would create a physically unified continent of Atlantropa, which would have an exploitable landmass that would make it competitive with Asia and America. This was important, argued Sörgel, for providing Europeans with the breathing space and the agricultural land they needed to form a new postnational unified political structure.[31]

In addition to the benefits of bounded bodies of water, Sörgel and Siegwart hoped that the man-made lakes would increase evaporation and thus precipitation. This, Sörgel argued, would gradually moderate the climate in Africa and allow for a "healthier life" for the anticipated European settlers. Once again, Sörgel copied Roudaire, who had expected a similar effect with his *mer intérieure*. In the early twentieth century, the proposition that large bodies of water changed the local climate had become generally accepted among engineers after longer-term observations around large man-made reservoirs. In 1940, the Hungarian Sahara researcher, motorist, Nazi spy, and real-life inspiration for the protagonist of *The English Patient* Count László Almásy announced his

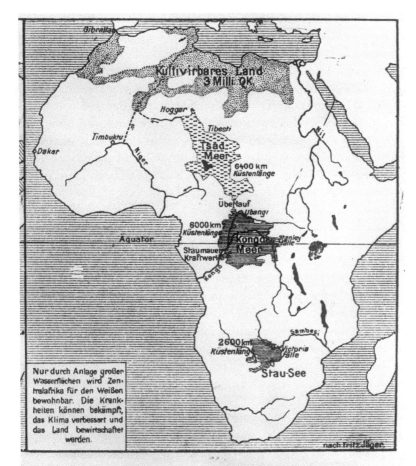

Erschließung durch Bildung von Binnenmeeren.

FIGURE 4.5. "Development through Inland Seas" showing the projected Chad,
Congo, and Zambezi seas. The caption reads: "Only through the creation of large
water surfaces will Africa be rendered inhabitable for the white man. Diseases
can be fought, the climate can be improved, and the land can be cultivated."
Source: Archives of the Deutsches Museum in Munich.

support for Sörgel's plans to create an inland sea in northern Africa. Like Sörgel himself, he was convinced that introducing a body of water would increase
precipitation and raise the groundwater table under the Sahara. And this transformation of the Sahara, Almásy added in somewhat awkward prose, was "one
of the enormous tasks that await the economic new order of human ingenuity
and enthusiasm [*Schaffensfreude*] in the coming years."[32]

Almásy's argument mirrored the rationale that Roudaire and his supporters had used in the 1870s and 1880s. After the French engineer's death, the idea of engineered climate change had survived, periodically reappeared in similar projects, and become part of the standard repertoire in the literature on deserts. A 1925 article published in a German popular science magazine called deserts "the biggest heat squanderers" on earth and mentioned the need for big evaporation surfaces of hundreds of thousands of square kilometers to bring rain to desert valleys, and especially the surrounding mountains. If reservoirs changed the climatic conditions in their immediate surroundings, the reasoning went, then very large reservoirs could change the climate of far larger areas. To those who believed that technology used on a grand scale could change atmospheric phenomena and macro-climatic—or even global climatic—conditions, this offered a whole new outlook on what meteorology would include in the future: active climate transformation seemed to have moved into the realm of the possible. A 1927 book on this subject called weather manipulation "the supreme issue of practical meteorology." The technological tools available to engineers now allowed Europeans to imagine modifying and moderating extreme climatic conditions, like those found in deserts or tropical forests. And "practical meteorology" included Europe and its climates as well as environments beyond the continent as potential targets of improvement projects.[33]

Sörgel made use of this new field of applied meteorology in his propaganda for Atlantropa. In an aside in his first major publication on the subject, he cited Paul Sokolowski's *Die Versandung Europas*, or "The Sanding-Up of Europe," from 1929. In this book, a confused mix of proto-environmental, anti-Bolshevik, anti-Semitic, anti-industrial, and eugenicist ideas, Sokolowski saw Europe threatened by desertification from the east. This process, Sokolowski seemed to suggest, was both a cultural and an environmental process, based on the effects of the "mobilization and socialization" of the soil under both capitalist and communist rule. What Sokolowski was trying to argue in detail remains obscured by his inscrutable prose, but his pessimism was clear enough: desertification was progressing and the time for halting it had already passed. Sörgel was not entirely persuaded by the book, feeling that Sokolowski had overstated his case. Obviously, the defeatist tone of the book was not in line with Sörgel's own revolutionary optimism. Sörgel did concede, however, that Sokolowski had hit a nerve. In language just as opaque as Sokolowski's, Sörgel wrote that the "sanding up of Europe by Russia" was "significant in the synthetic context of the entire course of worldly events."[34]

While Sörgel did not clarify his statement further, he was clearly concerned about climatic conditions and environmental changes in Europe. In his first article on Atlantropa, he already referred to the beneficial effects that the project would have on the reclaimed land around the Mediterranean basin.

Without giving much explanation of the causal connections between the engineering work and attendant climatic changes, Sörgel hoped for temperate climates from Italy to the Balkans. In the 1932 book *Atlantropa*, Sörgel went into more detail. He tried to allay the fear that damming the Mediterranean would change the climate for the worse in Europe. This was a response to criticism that lowering the water level would lead to an increase in atmospheric pressure and, ultimately, to an unpredictable shift in the high-pressure areas over the Azores that significantly influenced the European climate.[35]

Sörgel contended not only that the change in atmospheric pressure would probably be too small to have any discernible effect but also that—in the unlikely case that pressure areas were to shift—the change would have positive effects on the climate of Northern Europe, which he deemed to be "very much in need of improvement." Moreover, Sörgel wrote, the dam at Gibraltar would actually help to increase the power of the Gulf Stream. He reasoned that the concrete barrier between Spain and Morocco would cut off a cold undercurrent flowing from the Mediterranean to the Atlantic, which was counteracting the stream of warm water from the Gulf of Mexico. Without this undercurrent, the Gulf Stream would speed up and flow on further to the east, thus making Northern Europe warmer. In a newspaper article the following year, Sörgel expanded on this idea: Europe was currently in the process of cooling down. "If we want to have any sun at all or experience a spring with blooming trees," Sörgel warned, "we have to move southwards." There was one other possible solution, however: with the help of modern technology, Sörgel wrote, Northern Europe's climate could be improved, and the Gulf Stream could be shifted back to its supposedly original course.[36]

Once again, Sörgel argued that this would restore the environment to its "original state." This time, however, the intervention would not counteract the ongoing desiccation and warming of the climate, as in northern Africa, but offset the supposed cooling of the European climate. Sörgel presented Atlantropa as a remedy for both climatic extremes: it could transform hot deserts into green landscapes and the cold Northern European zone into warmer climes. For a while Sörgel seemed to have lost faith in the latter part of this design, or at least in the mechanism of engineering the European climate that Siegwart described in a letter from 1934. In that instance, Sörgel called Siegwart's reasoning a "fallacy," but it did not keep him from devoting a number of pages to exactly this topic in the 1938 book on Atlantropa. Illustrated with strangely whimsical maps of the Atlantic Ocean starring Poseidon, the chapter repeated the argument that the Gulf Stream could flow unimpeded after the Gibraltar dam was built, bringing with it warm and moist air to Europe (Figure 4.6). And this engineered alteration, Sörgel wrote, would counteract the cooling of the continent, as well as moderate the aridity.[37]

Abb. 11. **Der Golfstrom, wie er heute verläuft.**

Aus dem Mittelmeer kommt eine kalte Unterströmung (a), die wie ein Abwehr=
polster wirkt und den Golfstrom nach Norden und Süden ablenkt.

Abb. 12. **Der Golfstrom, wie er durch den Gibraltardamm seinen Lauf
ändern würde.**

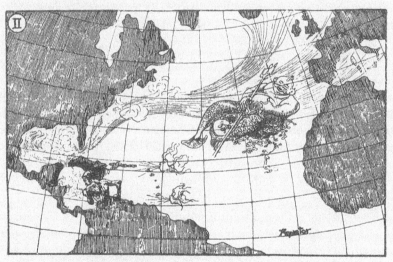

Wenn man die Straße von Gibraltar durch einen Damm sperrt, folgt der Golf=
strom seinem natürlichen Lauf. Europa, das einer gewissen Versteppung aus=
gesetzt ist, bekommt dann wieder feuchtwärmeres Klima.

FIGURE 4.6. Herman Sörgel's visual representation of the effects of the
Gibraltar dam on the course of the Gulf Stream (second panel). *Source:* Herman Sörgel,
Die drei großen "A": Großdeutschland und italienisches Imperium, die Pfeiler Atlantropas
(Munich: Piloty & Loehle, 1938), 25.

Next to concerns over renewable energy, the climatic benefits of Atlantropa became a central theme in Sörgel's writings. This was not only because of the architect's inclination to envision engineering environments on a big scale, but also due to his focus on what he perceived to be the issue of nation-states. Sörgel conceived of Atlantropa as a means to overcome the division of Europe. Climate, or rather climate change and desertification, offered challenges that required thinking, structures, and institutions that transcended the nation. After all, climate change did not care for national borders and could affect entire regions or continents. Sörgel believed that any attempt to counteract detrimental desertification, as well as effect beneficial climate change, would require a unified European response.

It was this political ambition of Atlantropa, above all else, that seemed suspect to many contemporary commentators. The nationalistic atmosphere of Europe after the First World War was not receptive to visions of cooperation and unity. And thus newspaper articles at the time often echoed Sörgel's technological enthusiasm while discounting his political plans as utopian. In a first reflection on Atlantropa in 1929, an article in the *Neuköllner Tageblatt* described the idea of a dam at Gibraltar as "not even all that absurd," though the author could not envision how the project would ever be able to find unified political support: "Here, the technological project becomes a true utopia."[38] And yet rather than "just" revolutionizing the political landscape of Europe and Africa, Sörgel had even further-reaching plans. His hope for a climatic overhaul of both continents always worked on a cultural level as well. In Sörgel's plans, climate engineering thus became the linchpin in creating not just new landscapes fit for European colonies and new political structures to govern them but also an entirely new society or civilization, free of the ills of environmental decline and modern civilization.

5

Europe's Last Hope

ACTIVE GEOPOLITICS AND CULTURAL DECLINE

HERMAN SÖRGEL was always cognizant that the claims he made about Atlantropa's potential to change climates, allow humankind to take over more land, and ultimately save European civilization might sound overly optimistic or even outright fantastical. In his articles, essays, and books he attempted to counteract the critique of impracticality. Atlantropa was surely a project on an unprecedented scale, Sörgel conceded, but that did not mean that it was an unrealistic dream: technology had advanced far enough to make Atlantropa a viable option. More than that, he argued with a normative flourish, the hydrostatic conditions of the Mediterranean "virtually *demanded* a systematic exploitation, a development on the basis of the already existent technological possibilities."[1]

Sörgel invested modern technology with an almost mythical power to overcome the problems of modernity, which he saw as a failure of politics to seize newly available instruments capable of tackling environmental, social, and economic issues that together would spell doom for occidental civilization in the not-too-distant future. While Atlantropa remained an outlier in its macro-engineering ambitions on a continental scale, it represented just one among the many extreme responses to the crisis of the interwar period in Europe. The project also struck a chord among cultural critics who saw connections between civilizational and environmental decay and for whom climate was not merely a static phenomenon but both an indicator of decline and a potential target for amelioration and regeneration.

In the 1930s, and especially during the Third Reich, Sörgel moved Atlantropa away from his original idea of a peaceful pan-European undertaking. He now started to fit his plans into the predominant fascist *Weltanschauung*, referring to the Berlin-Rome axis as the center of Atlantropa and highlighting the colonial and expansionist dimension of the project. At the same time, Sörgel further developed his ideas of a synchronous and interrelated cultural and environmental decline of Europe, which reflected and refracted some of the conservative-völkisch ideology of the day, mixing philosophical with climatological concepts. He also drew heavily on the tenets and terms of geopolitics but used climate engineering to propose an inversion: rather than the given geography determining the balance of power between nations, humans could now transform geographical features to shape power dynamics, and thus cultural developments, in the direction they desired. This would work particularly well in places that had not been settled by white Europeans and that could, according to both the racist-imperial convictions of Sörgel and fascist ideologies in general, be transformed into new environmental and cultural edens.

Blank Slate

From the early days of Atlantropa, Sörgel sought to present his project in large exhibitions to gather popular support. He targeted world expositions and other large international events and finally had the idea for an Atlantropa World Exhibition, which he pursued and planned with unrelenting tenacity—even through Germany's years of international isolation after the Nazi rise to power. While these grand plans ultimately failed, Sörgel toured Germany with smaller exhibitions in the early 1930s.[2]

During the crisis years of 1931 and 1932, he presented Atlantropa to the public in various German cities. The exhibition tour was not a complete failure—as newspapers articles from the time attest—but the enthusiastic reception that Sörgel had hoped for failed to materialize. In the following years, the German architect thought about all possible types of media to propagate his project. He played with the idea of commissioning a feature film involving romance, intrigue, a dashingly bold engineer, and of course the dam at Gibraltar as the breathtaking setting. In fact Sörgel himself started to write the screenplay. He also wrote flowery poems about the great virtues of the project and the future society they would help to create. And Sörgel even outlined the structure for a future Atlantropa Symphony, musically exploring the construction of the Gibraltar dam and the cultivation of the desert, with a final movement in *allegro vivace* annotated with "deliverance through power—peace and joy, choir with organ."[3]

Perhaps less unusual, and ultimately more effective, was Sörgel's attempt to win over some of the most famous architects of the Weimar period. Peter Behrens, who had collaborated with Le Corbusier, Walter Gropius, and Mies van der Rohe in Berlin, commissioned his students in Vienna to design urban plans and buildings for the cities around the Mediterranean that would have to be moved and rebuilt after the projected shift of the coastline (Figure 5.1).

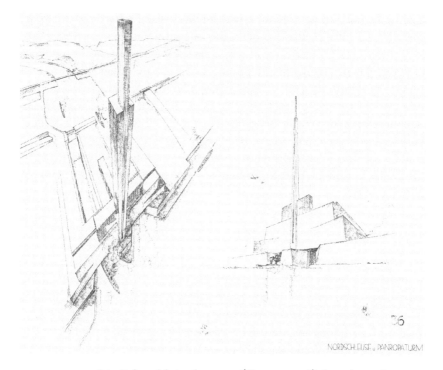

FIGURE 5.1. Peter Behrens' design for a tower ("Panropaturm") above the northern lock at the Gibraltar dam. *Source*: Archives of the Deutsches Museum in Munich.

Emil Fahrenkamp, a professor and later the director of the Kunstakademie in Düsseldorf, assigned the drawing of plans for a post-Atlantropan Gibraltar to his students. Lois Welzenbacher, Hans Döllgast, Fritz Höger, and other German and Austrian architects all spared some of their time to design different parts of Sörgel's project, such as power plants, canals, and dams. The Jewish-German expressionist architect Erich Mendelsohn, best known today for the Einstein Tower in Potsdam, volunteered to work on plans for a redesigned and enlarged Palestine and publicly advertised Sörgel's project.[4]

Why so many leading German architects became fascinated with Atlantropa is difficult to answer conclusively, but Sörgel's plans for land reclamation offered them something that no existing project or space could rival: a supposedly blank slate providing the unique opportunity to design without any restraints of existing structures or geographic obstacles. Moreover, Atlantropa also gave the architects a welcome distraction from the dismal circumstances of their professional lives during the worldwide economic crisis in the late 1920s and early 1930s. Despite its scale, many also regarded Atlantropa as an audacious yet not impossible scheme. After all, Sörgel emphasized repeatedly that the project's structures could be built with existing technologies.

Right from the first publications on Atlantropa, the tenor in the news was somewhere between benevolent tolerance and utopian enthusiasm, with some concerns about the political feasibility and desirability of the plans. In the 1930s, the general thrust of comments on Atlantropa did not change markedly, although the newest addition of the project—the African inland lakes—now drew the largest share of attention. A 1936 article by Richard Hennig—yet another contributor in the Atlantis debate of the 1920s and a prolific author on geography and technology—wrote quite favorably of the expanded Atlantropa project, but pointed out the inconsistency in a plan that decried overpopulation yet projected a decrease of the African landmass through the submersion of wide swaths of farmland and rich forests around the anticipated Congo Lake. Moreover, Henning argued, the expected surge of rain from the large inland lakes would further harm the already overly wet Congo area.[5]

While advancing an idiosyncratic argument, Henning's article still reflected one of the central trends in criticism of Atlantropa: commentators usually saw the greatest obstacle to the project's realization not in technological but in political and social challenges—be they colonial power struggles in Africa or the prospect of European unification. Even commentators who called for caution usually did not question the basic viability of the project. Hennig, for his part, described Sörgel's engineered lakes as an undertaking that would be "technologically feasible at not too exorbitant a price, and that one will nevertheless still eschew for the time being, because the induced changes would be of an all too radical kind."[6]

While the criticism of Sörgel's project resembled that of Roudaire's plans fifty years earlier, commentators found even less fault with Atlantropa's technological ambition. In part, Roudaire's Sahara Sea suffered from its own success: critics focused more on its practical shortcomings because the plans had already been received and discussed by various parts of the national and colonial administrations. Atlantropa never reached the same level of serious consideration among political decision makers, but there seemed to be something else at work as well: the widespread belief in the boundless progress of technology had increased even further since the days of Roudaire. And while the First World War had shown the horrors of industrial technology, it had also added to its clout and mystique, so that even potentially disinterested commentators took a relatively benevolent stance toward Atlantropa's feasibility.

In a 1929 forum in the magazine *Reclams Universum*, the eminent geopolitician Karl Haushofer commented in two paragraphs on Sörgel's plans, identifying overwhelming obstacles to the realization of Atlantropa. Haushofer expected a detrimental climate change in southern Germany from long-range atmospheric effects of the projected geographic alterations. He also feared a shift of the geopolitical center away from Central Europe and expected insurmountable resistance from coastal countries around the Mediterranean, which were bound to lose their ports along with their "entire littoral culture" with the coming of Atlantropa. The criticism, however, was once again concerned not with the project's technological feasibility but with potential unintended consequences and detrimental geopolitical effects.[7]

The shift of political and economic power from Central Europe to the Mediterranean region that Haushofer criticized was, however, an *intended* consequence of Atlantropa. Sörgel was convinced that only an expansion of Europe beyond its original borders could save the Occident. Once the European center of power was firmly established in the Mediterranean region, he believed, the areas around the downsized ocean would naturally come under European control and provide new living space for white settlement and agriculture. Ironically, this line of argument owed a good deal to none other than Karl Haushofer himself.

Active Geopolitics

Haushofer commented on Atlantropa again two years after his initial remarks. He praised the many improvements in the plans since the first publication of Sörgel's ideas but still did not sound convinced. Thinking on a large scale was certainly commendable, Haushofer wrote, but Europe "should not let itself be blinded, not even by beautiful plans and alluring designs." Maybe it was this renewed criticism that led Sörgel to call Haushofer a *Pantoffelritter* (literally, a "slipper knight"; figuratively, a "coward") in his personal notes.[8]

Despite this reproach, Sörgel borrowed heavily from geopolitics, which experienced a boom in Germany over the interwar years. Geopolitics, a theory emphasizing the role of geography and space in political power dynamics, had become a popular model as a way both to attribute the German defeat in the First World War to forces beyond individual human control and to identify ways out of the post-Versailles crisis. Despite this success, however, it did not manage to establish itself as a scholarly discipline in Germany. From the outside, critics attacked geopolitics for its lack of a sound method; from the inside, Haushofer did little to counteract the critique and ultimately even assailed the field that he helped create. As the leading geopolitician in Europe, he avowed that the subject was missing a solid framework but declined to answer the calls for a theoretical corroboration. The terms and ideas of geopolitics nevertheless remained important in German public discourse. In *Atlantropa*, Sörgel quoted the British geopolitician James Fairgrieve in support of the notion that the

Sahara had once enjoyed a wetter climate. With his ruminations about harvesting solar energy in the Sahara, Fairgrieve may have also been important to Sörgel's idea of the desert as a vital source of energy for the future, although the German architect focused on hydropower instead.[9]

Beyond the geopolitical focus on energy supplies, Sörgel was especially interested in the idea of Lebensraum, or "living space," that had been developed by the Leipzig geographer Friedrich Ratzel in the 1890s and later found its way into geopolitical discourse. In the first paragraph of his 1901 essay on Lebensraum, Ratzel laid out the main tenet of his theory:[10] "The conditions of all of life's development are marked by a great telluric [earthbound] element. Even if particular cases seem to be defined by local conditions alone: once we dig deeper, we find roots that are intertwined with the basic characteristics of the planet."[11] The content of this passage does not seem too far from the basic principle of current-day environmental history. And indeed Ratzel's approach was new and noteworthy in his time. His major work from 1882, Anthropo-Geographie, included a sweeping critique of various philosophies of history and, particularly, a full-frontal attack against idealism as a mode of historical explanation:[12] "The degeneration of the concept of history . . . is most conspicuous in Hegel, who, in a frequently cited dictum, only regards as history that 'which constitutes a substantial epoch in the development of spirit.' . . . How un-geographic these ideas are, how they show absolutely nothing of the expansion of the horizon which is always a necessary consequence of geographic studies, and what level of unjust blindness towards the nature of things they betray!"[13]

For Ratzel it was not the intangible "spirit" that ultimately determined historical development but migration, which he posited as "the fundamental theory of world history." The movement of humans had its necessary corollary in space. According to the Ratzelian anthropogeographical model of the early 1890s, the limited inhabitable surface area of the earth led to the competition and confrontation between different groups of people. Influenced by Charles Darwin's and Ernst Haeckel's theories of evolution, Ratzel conceived the resulting "struggle for room" as a deliberate modification of the "struggle for existence," a concept he saw as far too abstracted from geographical realities. The term Lebensraum did not feature centrally in early anthropogeographical thinking. When Ratzel used it in the early 1890s, it was a fairly innocuous concept, denoting the geographic area in which the life of a group of human beings unfolded within the limits of the particular environmental circumstances. In fact, Ratzel himself never provided a clear definition of Lebensraum in his work, a circumstance that was probably at least to some extent responsible for the success and longevity of the malleable term.[14]

As one among many, Sörgel adopted the concept but added an idiosyncratic touch. In his interpretation of Ratzel, the size or quantity of space, and not just its geographical and geological characteristics, determined its quality: "Expansive space is life-sustaining; in confined space, the struggle grows desperate." What this dictum referred to in terms of Atlantropa was clear: Europe, overpopulated and cramped, had to expand into the vast space of Africa in order to maintain its cultural authority and ensure its political survival. Sörgel also developed Ratzel's thought further and added *Kraft*, power or energy, to the model. According to the German architect and in line with geopolitical thinking of the time, energy *and* space determined historical development. Only with sufficient stores of energy could Europeans undertake the gigantic task of expanding their borders into Africa.[15]

Sörgel never understood geopolitics to be pure theory. He was convinced that the capabilities of technology had increased to the point that environmental conditions, like terrain and climate, would no longer have to be taken as an unalterable given. Sörgel developed an approach that could be called active geopolitics, in which technology's role was to shape geographical and geological conditions to "create a previously conceived political desideratum." With modern technology as a powerful ally, man no longer had to be dominated by nature. Sörgel was clearly not fazed by the sporadic calls from some of his contemporaries to be cautious about technology's potential. Sörgel held no such reservations: "What is possible technologically," he wrote, "has to be exploited economically."[16]

Technology over Everything

Sörgel's boundless confidence in Atlantropa often bordered on the arrogant, naïve, and obtuse. In his writings, he drew on all possible sources of support, including esoteric notions of cultural development and a belief in mystic forces controlling nature. Sörgel saw his plans for a new Atlantropan culture "in harmony with the world horoscope" and standing in the sign of the *amphora* (Aquarius) in his own "cycle of world religions." He also repeatedly referred to the "imponderable processes of nature" and cited "unfathomable cycles" that, in his interpretation, were beyond human understanding and control. "Nature," he wrote, "is more powerful than man."[17]

This did not mean, however, that nature was beyond the scope of human intervention. Above all else, Sörgel believed in the power of technology and its ability to transform the environment. In his view, the "imponderable processes" actually worked in favor of his Atlantropa plans because they would eventually recalibrate any possible imbalances of nature that technological interventions could cause in the short term: thus, an uncharted water cycle would balance out the dislocation of water from the Mediterranean and prevent the earth's axis from shifting, and atmospheric processes beyond current human understanding would prevent climatic deterioration in Europe. In Sörgel's thinking, technology even played a role similar to that of these natural processes: while it could lead to unintended side effects, it also had the power to repair the environment and restore it to equilibrium. Rather than transforming nature to a new and unprecedented state, technology in the Sörgelian sense could in fact help to restore the original, authentic state of nature that had been lost through geological disasters, unchecked population growth, and anthropogenic damage like deforestation and desertification. In his pursuit of Atlantropa, Sörgel sought to recultivate the once green and fertile Sahara but also to reconvert the Mediterranean to the lake it had been long ago, before water broke through the Strait of Gibraltar, causing a gigantic flood.[18]

Technology, according to Sörgel, was in no way opposed to nature. It was a tool in the hands of engineers, who could use it to repair environmental damage and look for possible improvements in line with natural preconditions. Describing the projected reservoirs and power stations in the area of the Adriatic Sea, Sörgel emphasized that Atlantropa was simply following a path laid out by nature itself: "With the Dalmatian Islands, nature has already built the elongated dams required for reservoirs. Man only has to finish this work in the spirit of nature." Despite the enormous geological and climatic changes that Atlantropa would bring about, Sörgel never saw his project as radical or violent. The engineer's task was not to force his will on nature but rather to use hidden natural potential. In practice this meant carefully analyzing the physical world, searching for evidence of prior environmental conditions and favorable natural features, and finally applying modern technology to transform the earth according to human designs, but within the limits of natural precedent. In Sörgel's own words: "The engineer can 'correct' nature."[19]

While these statements might suggest that Sörgel held a rather cautious and limited view of the power of technology, the sheer scale of Atlantropa alone makes it obvious that that was not the case. Sörgel argued for a use of technology that heeded certain natural conditions and processes, but he did not seem to perceive any innate limits to technology's reach. When confronted with questions about the viability of the project, Sörgel simply referred to the incessant march of technological progress that would take care of the problem. When his collaborator Siegwart asked about the necessity of a protective weir in front of the dam at Gibraltar to prevent possible geological disasters or military attacks that could cause a catastrophic flood of the reclaimed land in the Mediterranean basin, Sörgel seemed unfazed. He replied that a yet undiscovered construction material would render the protective dam superfluous. In publications, Sörgel also repeated remarks on future solutions to technological problems of the project, be they the energy-efficient desalinization of seawater for irrigation or the construction of further dams around the

Mediterranean. Whenever these obstacles became pertinent, Sörgel argued, technological development would undoubtedly catch up and present a solution.[20]

Sörgel saw technology as an independent force that developed irrespectively of human intentions and had progressed quicker than social change. "Politically, we still live in a Europe that, in its basic outlines, took shape in the age of the horse-drawn carriage," he wrote in 1932, emphasizing the internal contradiction between old political structures and the technological marvels of the twentieth century. Sörgel penned these lines when the Weimar Republic was teetering on the brink of collapse. The effects of the Great Depression were felt deeply in a Germany that had never quite recovered from the First World War. Radical political parties on the right and the left were on the rise, and the moderate chancellor Heinrich Brüning of the Catholic Center Party presided over a highly unpopular policy of severe fiscal austerity. Behind the scenes, powerful political factions on the right already schemed to bring down Brüning's minority government, which ultimately toppled under the pressure in May 1932. The sense of an all-encompassing crisis pervaded economic and political debates in the waning years of Weimar democracy.[21]

Especially among engineers, Sörgel's view that political realities had not kept pace with technological developments in Europe was widespread. While Sörgel tried hard to stay above the fray of day-to-day politics, he explicitly agreed with the calls for some form of technological rationality to take the reins. It was technology—so often abused—that had led humans toward "higher perfection," he argued in a magazine article on engineering: "Not in a battle *against* technology and machines, but only in union *with* them, can we progress." Ultimately, engineers would not just modify environmental and geographic conditions but also provide the ground for the rise of a new culture or civilization, finally in tune with technological progress.[22]

Sörgel's beliefs showed marked similarities to the technocracy movement, which flourished for a very short period in the early 1930s. Technocratic tenets, centered on replacing politicians with scientific experts and monetary valuation with a kind of economic energetics, found a particularly receptive climate in the United States after the onset of the world economic crisis. Once the movement made it over to a similarly hard-hit Germany, Sörgel found a congenial set of ideas that complemented and supported his plans for technologically reorganizing Europe. He quickly became involved in the movement himself, writing the foreword for the German version of Wayne W. Parrish's main oeuvre, *An Outline of Technocracy*. In his book, Parrish emphasized that civilization was built upon energy and that any attempt to understand the structure of the economy had to take this fact into account: only energy could measure labor and value objectively. The foreword was an enthusiastic endorsement. Sörgel fully agreed with the emphasis on energy and supported the core demands of the technocratic movement. "It is not technology that has to adapt to the economic system," he wrote, "but vice versa: we have to find a new societal order for the factual results of machine industry." Sörgel's deprecatory view of the political process was fully in line with the technocratic movement overseas, but it was also the product of a crisis at home. The global economic depression and the continued instability of the Weimar governments spurred the search for novel alternatives, far removed from the bleak realities of interwar Europe. Atlantropa can be seen as one of these alternatives—a "utopia of the crisis," as Alexander Gall has aptly called it in his study of Sörgel's project.[23]

The Decline of the West

The feeling that something was fundamentally wrong with modern society was certainly older than the Great Depression and the last convulsions of a crumbling Weimar democracy. The most popular prophet of this cultural pessimism during the interwar years in Germany was the philosopher Oswald Spengler, whose *Decline of the West*, first published in 1918, became hugely influential for a whole generation of German thinkers. Spengler's vision of history was one of inevitable catastrophe. Cultures, like human beings, had a life cycle: they grew up, became powerful, and then waned until they died and vanished. Europe had already reached the phase of old age, and the *Untergang*, the demise, was near. Sörgel was an avid reader of Spengler's theories and incorporated his ideas and terms into the literature on Atlantropa. He referred to Spengler's oft-cited *Raubtier Mensch*, or "human beast of prey," in describing the destructive tendencies of mankind, illustrated all too well by the First World War and what Sörgel described as the "cage-like" condition of national borders in Europe. Even the motto of Atlantropa—"either decline of the occident or Atlantropa as a turning point and new goal"—was an homage to the most famous work by the philosopher.[24]

Sörgel, however, had an idiosyncratic way of reading and understanding Spengler's works, which were often elliptical and open to interpretation. Spengler's 1931 publication *Man and Technology*, his only half-successful attempt at clarifying some of his more obscure lines of thought, still contains one of the more unambiguous, and deeply pessimistic, statements on technology: "The struggle against nature is hopeless and yet it will be waged until the end." There could hardly be an assertion more opposed to Sörgel's ardent belief in technology's potential to transform nature and culture. Spengler and Sörgel also evaluated a technological vision of the world differently. Spengler lamented the rationalistic and utilitarian view of nature, criticizing the gaze of the modern man, which saw a waterfall only as a potential source of energy. But what the philosopher regarded as an ominous sign of cultural decline, the architect embraced with enthusiasm as a sign of a new age: Sörgel touted the vision of the Mediterranean Sea as a power station as the heart of an emerging Atlantropan society rather than the last vestige of a declining European culture. The intellectual distance between the two men was also clearly apparent in their respective views of environmental change: Spengler saw "changes in climate that threaten the agriculture of entire populations" as a deleterious effect of the ascendancy of technology, while Sörgel acknowledged anthropogenic desertification but sought to fight and reverse it through technology.[25]

Sörgel countered Spengler's famous dictum "optimism is cowardice" with an unwavering belief in technology's potential to save humanity. Sörgel contended very confidently that Europe could be transformed into a new cultural and political entity through large-scale engineering and the production of enormous quantities of energy. To be sure, Spengler also had his own more optimistic moments, especially during his half-hearted attempts to synchronize his ideas with the political project of the reactionary right in Germany. This politicized version of his outlook for occidental culture, however, quickly became embroiled in the völkish ideas of his most avid right-wing readers. Sörgel, in contrast, initially stuck to a pan-European model of technological development. In fact, the cooperation of European states was an important precondition for the success of the gigantic engineering structures of Atlantropa. In his vision for the project, Sörgel never attempted to conserve or strengthen any existing cultural and political structures, while he nevertheless borrowed heavily from colonial, racist, and technocratic ideologies.[26]

Despite their different views on technology, Sörgel attempted to fit Atlantropa into Spengler's idea of cultural life cycles, emphasizing once again that his project would found a "new culture" in opposition to America, which he saw undergoing the same process of decline as Europe. Sörgel nevertheless did not become an outright Spenglerian. In the foreword to Parrish's book, Sörgel distanced himself explicitly from Spengler. Europe's demise was not so much an ideological issue as presented in Spengler's writings, Sörgel cautioned, but a question of technology in practice.[27]

The philosophical differences between Spengler and Sörgel also point to a broader phenomenon. European responses to technology in the first half of the twentieth century are, in general, difficult to classify. The common fear of mechanization and an "un-organic" utilitarian rationality often combined with the hope that technology could solve the daunting issues of overpopulation, energy production, and unemployment. The situation was made even more opaque by the fact that different attitudes toward technology did not map neatly onto political alignments. There were no typical responses on either the left or the right. The two sides, in fact, had in common the often-voiced goal of creating a "New

World" or a "New Man." The radical social and political transformations of the time called for even more radical comprehensive solutions.[28]

In this sense and despite his idiosyncratic pan-European beliefs, Sörgel's faith in the rise of a new society was not out of the ordinary. But before this new society could emerge, he argued, there had to be enough space available for modern technology's power to unfold. The Sahara, in Sörgel's words the "vacuum" in front of Europe's gates, was the ideal proving ground for engineers to deliver on this promise. Transforming the climate and cultivating one of the largest and most inhospitable deserts of the world would conclusively prove the power of modern engineering and the reality of active geopolitics. A 1940 article in the journal of the Verein deutscher Ingenieure, the largest association of engineers in Germany, portrayed the desert as the last stand against modern technology. The task at hand was to overcome this "zone of resistance to technology." The ultimate goal was not necessarily the utter domination but the technological understanding and control of the most hostile landscapes on earth. In the ever-expanding reach of rational planning that the Jewish-Hungarian sociologist Karl Mannheim described as a distinctive feature of the twentieth century, the desert would eventually become a "functional part of the social process," or—in other words—just another landscape made to serve human needs. This would also mark the ultimate success of the engineer and, in Sörgel's plans, the realization of Atlantropa.[29]

South versus East

Herman Sörgel had only four years to publicize his project without fear of censorship. Hitler's accession to power in 1933 threw Sörgel's plans into turmoil. The self-professed adherent of a European union had to be wary of the new government, which was openly nationalistic and bellicose. However, Sörgel accommodated and adapted. He had always been more an "Atlantropist" than a firm believer in any political ideology. While Sörgel actively promoted European cooperation and collaboration, he also tried to gather the support of fascist Italy in the late 1920s. In response, Mussolini informed Sörgel through the Italian consulate that he was following the project with "great interest," although he never seems to have followed up on this note. Once the Nazis came to power in Germany, Sörgel—like so many of his compatriots—laid low for a while and then revealed a great knack for opportunism.[30]

In a 1936 newspaper article, Sörgel published his thoughts on the new regime and tried to make Atlantropa compatible with National Socialist ideology. The architect now cited the Nazi ideologist Alfred Rosenberg's *Myth of the 20th Century* in support of his theories on the rise of "a new cultural epoch of the white man." Sörgel visibly struggled with the new primacy of racial distinctions in Nazi Germany, arguing somewhat confusedly that a race-based order in Europe could be realized only once an ill-defined international body would conquer new *Lebensraum* in northern Africa. In another article published two years before, Sörgel avowed that he would now rescind his earlier ideas of an "Atlantropa Party" and explicitly wrote about his intention to square his ideas with the "current mode of thinking [*Zeitströmung*]." What Atlantropa and the Third Reich had in common, he argued, was the attempt to break interest-driven economics and Roman land rights. Sörgel emphasized the corporatist and collective dimension of Atlantropa, which overlapped somewhat with Nazi dogma.[31]

In fact, Sörgel's thinking complemented Nazi ideology in other ways. Despite his unwillingness to get involved in party politics, the German architect displayed antidemocratic and authoritarian tendencies from the earliest days of his propaganda efforts. In an undated note, he described "so-called democracy" as "the opposite of general human rights." When faced with the question of how to procure labor for the construction of the massive dam at Gibraltar, Sörgel immediately proposed that convicts should undertake the arduous and potentially dangerous work. Atlantropa was, moreover, always designed as a project of white imperial domination over Africans. Sörgel openly expressed racist notions and wrote about submerging a great part of the Congo as a blessing, as it would exclusively harm the indigenous population. Without the project, Sörgel argued, "[Africans] would increase in numbers, until they would finally eat everything the land can produce and not export anything anymore."[32]

Sörgel's collaborator Siegwart expressed similar views in their letter exchanges and put a peculiar spin on the projected engineered climate change in central Africa. He wrote to Sörgel that for the sake of the continued domination of the white race it was imperative to destroy landscapes in which only Africans could survive. As an argument to persuade Europeans of the necessity of his vision, Sörgel framed it as a response to the threat posed by the "black race": the Black population of Africa, he claimed, had superior reproductive potential and a higher resistance to tropical diseases and would overpower the whites unless Europeans colonized and transformed Africa. "Throughout the entire history of the dark continent," he wrote, "the black [man] has proven incapable of higher cultural development and he will always remain this way." This mindset, widespread among European political circles at the time, was obviously consistent with Nazi notions of racial hierarchy.[33]

Sörgel's 1938 publication *Die drei großen "A"* ("The Three Great A's," i.e., Asia, America, and Atlantropa) represented a much more direct link to Nazi ideology and marked the culmination of Sörgel's effort to curry favor with the new German leadership. The subtitle of the book now highlighted "Greater

Germany and the Italian Empire" as "the pillars of Atlantropa." Using a quote from Hitler's *Mein Kampf* on the importance of setting the right goals in planning as its introduction, in *Die drei großen "A"* Sörgel voiced the hope that the two fascist countries would become "the womb of Atlantropa." In the book, Sörgel adopted Nazi sentiments in his explicit hostility to urbanization. While it was certainly difficult to portray Atlantropa as opposed to urban development— after all, the project included plans for new megacities around the Mediterranean—Sörgel argued that the reclaimed land would make for more organic and less crowded new cities. Rather than the Manhattan-like chaos of skyscrapers that Sörgel predicted in his 1938 book, the new Atlantropan cities would expand "naturally" and with space to spare.[34]

In *The Three Great "A's"* Sörgel also attempted to distance himself from the antiwar rhetoric that had marked Atlantropa propaganda in the late 1920s and early 1930s and to disassociate himself from the ideas of Richard Coudenhove-Kalergi, the founder of the pacifist Pan-European movement that Sörgel had still cited as a major, if flawed, source of inspiration in 1929. Sörgel highlighted the struggle and strength it would take to build Atlantropa. And yet this attempt to portray Atlantropa as more aggressive and warlike did not sound entirely authentic. In fact, Sörgel continued to think about the "United States of Europe" as the only way to evade war. And even in *Die drei großen "A,"* he still called for a "European union," if now under a German-Italian leadership rather than a more or less equal alliance of all large European states.[35]

While Sörgel's attempts to rephrase his Atlantropa propaganda to make it more appealing to the Nazi government may not have been entirely successful, the project did have something in common with Nazi approaches to economic planning. One of Atlantropa's most fundamental goals was to end European dependence on any essential resources from abroad. The Gibraltar dam, the Congo power stations, and the transformation of the Sahara into farmland were all conceived to avoid energy shortages, whether measured in horsepower or calories. This effort to reach independence—particularly in terms of the energy supply—was also a central point of strategies in the Third Reich—first to rebuild the German economy after the Great Depression and then in preparation for war. Autarky became the catchword of National Socialist economics. The concept had long been a staple of economic thought in Germany, but it became an especially popular topic of discussion during and after the supply crises of the First World War. Concurrently, autarky had become a central term in geopolitical theory, particularly in the writings of Rudolf Kjellén, the Swedish "father" of geopolitics. Nazi ideologists found in it a concept consistent with their ideas of national independence.[36]

Already in 1933, an economic treatise on the impending reorganization of the economic system under the Third Reich emphasized autarky, especially in terms of the food supply: "Autarky means independence from foreign lands," the book stated in almost Sörgelian language, "and thus leads to economic and political freedom. Consequently, even the last remaining potential has to be exploited, expanded, and employed." Self-sufficiency was particularly important, and just as hard to come by, in the energy sector. Under the guidance of Fritz Todt, who had overseen the development of the German Autobahn, Nazi Germany embarked on a program of hydropower exploitation in the Alps. As the Second World War was wearing on, and after the loss of oil fields in Russia and Romania and the destruction of synthetic oil plants, the Nazi leadership scrambled to find sources of fuel to keep the war machine running. Exemplary of this last-ditch effort was the strangely named Unternehmen Wüste, or Operation Desert, which aimed to produce shale oil in the Swabian Alps with slave labor from concentration camps.[37]

Sörgel's own notion of autarky always centered on hydropower, but his goals of energy independence interested the Nazi leadership. In fact, Sörgel was in contact with officials in the Reichsstelle für Raumordnung, a Nazi office entrusted with the rather nebulous task of ordering and planning the space of the German Reich. The Reichsstelle was aware of Sörgel's ideas and commissioned him in 1941 to investigate the energy supply of Europe and its spatial relations to Atlantropa, although nothing seems to have come of this proposed collaboration. There is no evidence that Sörgel ever worked on the project, nor that the Nazi bureaucracy showed any further interest. Sörgel's other contacts with the Nazi bureaucracy, such as his attempt to approach Richard Korherr from Heinrich Himmler's Reich Commissariat for the Strengthening of Germandom, remained superficial as well and did not progress beyond the exchange of pleasantries. In fact, Sörgel's efforts to ingratiate himself with the Nazi leadership never progressed beyond the initial stages. Ultimately, Atlantropa was just too far removed from Nazi plans for a new European order under German dominance.[38]

The main difference between the projects, however, was geographic. Sörgel, like Nazi planners, thought in unprecedented dimensions about environmental and climatic transformation. Also like Nazi planners, he referred to a "European eastern problem" and called for "a rampart against the east." But Sörgel was not interested in the "Asiatic steppes and marshes" that would become the focus of Nazi policies. He argued that Germany had to aim for a "place in the sun" and the expansion of territory and power in a north-south direction. During the Nazi period, Sörgel realized quickly that his focus on Africa as the future colonial territory was not in line with Nazi ideas. He tried to promote the advantages of his project, referring to the geopolitical importance of a north-south extension for food production and access to raw materials and thus for autarky. "West-east can never lead to self-sufficiency," Sörgel argued,

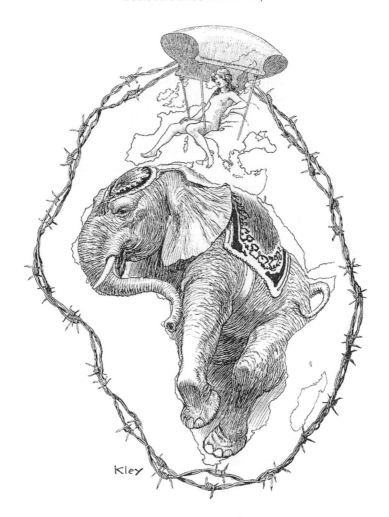

FIGURE 5.2. Drawing for Atlantropa propaganda by the German cartoonist
Heinrich Kley. *Source*: Archives of the Deutsches Museum in Munich.

"because it only covers a limited section of zones. North-south, however, is
identical with autarky, because it possesses lands in all zones of the earth."[39]

In his private notes and manuscripts, Sörgel voiced candid criticism of Nazi
plans in the East. He chastised eastward expansion as absurd, unnatural, and
ultimately futile. In a published article, he emphasized the danger of an Islamic
world empire if Europe did not actively claim and colonize Africa. He also
warned of a "yellow peril" from the East as early as 1931 but did not prescribe
the same efforts there. India, China, and Japan, Sörgel argued, were "hostile
to the [white] race." The solution was to counter Asia's rising influence with a

powerful union of Africa and Europe. Russia, the focal point of colonial policies under the Nazis, was another case altogether. Probably following some of Sokolowski's arguments, Sörgel did not see any hope of change or transformation. He described Russia as unable "to create culture [*Kulturschaffung*]" and argued that "for all intents and deepest purposes, Russia is and remains infertile, negative, depressed, nihilistic." The engineered transformations that would be able to create Atlantropa out of the sea and the desert, Sörgel seemed to contend, would be powerless against Russian nature.[40]

Sörgel's reservations obviously did not alter the Nazi plans. Even worse for Atlantropa, it seems that Sörgel fell entirely out of favor with the Nazi leadership in the early 1940s. Already in 1939, Sörgel was informed that support for the Atlantropa project from any part of the German bureaucracy was "completely impossible." Two years later, Nazi wartime censorship allowed Sörgel to publish, but only without political references or allusions to actual colonial plans of the German government. In the same year, Sörgel was unsuccessful at registering an Atlantropa Society in the Bavarian register of associations. Because of Nazi censorship, Siegwart and Sörgel had to reroute their letters through the Swiss consul in Munich and conceded that a free exchange of ideas was not possible under the given constraints. In 1942, things got even worse for Sörgel: the Propaganda Ministry prohibited the planned publication of *Kraft, Raum, Brot* ("Energy, Space, Bread"), another treatise on Atlantropa, and ultimately banned the author from circulating any more of his writings. A year later, the Gestapo proscribed any propaganda activity on behalf of Atlantropa. In an unpublished note, Sörgel complained about the "hostility, antagonism, and persecution" of the Atlantropa plans since 1933, although he still ended this grievance with a Hitler quote.[41]

Into Oblivion?

When the aerial attacks on Munich intensified and his house was bombed in 1943, Sörgel moved to the small mountain town of Oberstdorf. Here in the German Alps he sat out the end of hostilities and the ensuing chaos of defeat. Very quickly after the American invasion, Sörgel took up his propaganda activities once again. He used the freedom from Nazi censorship to bring the Atlantropa Institute into the open and from 1946 published the *Atlantropa Mitteilungen*, a newsletter promoting the project. Sörgel assembled a cast of oddball supporters for his project, who ranged from the Swiss author John Knittel to the autodidactic Faust scholar Karl Theens. With funds from his wife, Sörgel envisaged founding an Atlantropa University in the former Nazi cadre school in Sonthofen, but the plans quickly ran aground.[42]

True to his role as a political chameleon, Sörgel tried to take advantage of the new situation after the end of the war. He cooperated with the Allied forces stationed in Germany and denounced the Nazi regime. Sörgel hoped that the second world war in thirty years had finally brought about conducive conditions for a peaceful European unification but remained distrustful of democracy. When he called for an international government with representatives like Winston Churchill, Jan Christiaan Smuts, George Bernard Shaw, Thomas Mann, and Albert Einstein, he was sure to emphasize that their role would be to "demand and dictate" rather than "propose or plead." In further notes, Sörgel added others to his list of a "council of twelve" that would prepare the world for Atlantropa, including Pope Pius XII, US president Harry Truman, Soviet Union premier Joseph Stalin, Indian prime minister Jawaharlal Nehru, chairman of the Communist Party of China Mao Zedong, the philosopher José Ortega y Gasset, the postwar architect of European integration Robert Schuman, the American diplomat Ralph Bunche, and—less expected in a list of such outstanding personalities—the Austrian writer and journalist Anton Zischka.[43]

Zischka, who had served the Third Reich's propaganda efforts, developed theories that were clearly compatible with Sörgel's vision of the world. In 1938, Zischka had already published a book on what he regarded as an impending world hunger crisis. In the same book, he turned his view to the Sahara, borrowing Sörgel's argument that only water was needed to make the desert as fertile as it had been in classical times. Zischka even explicitly cited Sörgel's plans as a "great goal" for the future. After the war, Zischka allied more closely with Atlantropa, publishing in 1951 a book that proposed a combined "Eurafrica" as an escape from the destructive east-west tensions that had been building up since the end of the war. Once again mirroring Sörgel's rhetoric, Zischka characterized Africa as "the almost empty 'reserve continent'" and described the Mediterranean as a "lake" and a "hub" for Europe. Zischka emphasized the impending "climate war," in which the engineered Congo Lake would play an important role by counterbalancing similar projects of the Soviet Union.[44]

Here Zischka was probably referring to an assembly of plans by Stalin and the Soviet leadership, which proposed, among other things, rerouting rivers in Siberia and creating an inland sea the size of West Germany that would provide enormous amounts of energy and moderate the hostile climate in the region. The project was discussed in Germany in 1950, as some commentators feared that German prisoners of war would be used as the labor force for the construction. Sörgel used the plan's publicity to advertise his own project, highlighting the allegedly far superior features of Atlantropa. Conversely, the relevant authorities in the Soviet Union apparently knew of Sörgel's project as well and attempted to win the architect for their cause. This is at least what a newspaper reported in 1951, although Sörgel himself may have planted the information in order to increase pressure on the German government and the Allied powers to support Atlantropa.[45]

The support never materialized. The new world, marked by the rise of nuclear energy and a shaken European continent recovering from the Second World War, no longer seemed to have a place for Atlantropa. Sörgel struggled against this realization, but the publication of a short documentary film and a commemorative volume of his own Atlantropa poems to celebrate the twenty-fifth anniversary of his project marked the last big events in its history. The Atlantropa Institute, which had served mainly as a publishing house for propaganda, ran into deep financial trouble with the currency reform in West Germany in 1949. For unknown reasons, several members publicly announced their withdrawal from the board in 1950. In an article in the *Frankfurter Allgemeine Zeitung* in the same year, the geographer Kurt Hiehle sharpened his devastating critique of the project, which he had first targeted in 1948. The reclaimed land around the Mediterranean would remain as uncultivable as the African coastline, the irrigated areas in the Sahara would become vast salt lakes, and the smaller evaporation surface of the lowered Mediterranean might lead to detrimental climate changes, Hiehle argued. "The Gibraltar project is thus not a utopia that will bless mankind [*menschheitsbeglückend*]," he concluded, "but would actually cause incalculable damage."[46]

Around the same time, the official contacts that Baron von Loeffelholz, the in-house counsel of the Atlantropa Institute, had tried to establish finally collapsed. Neither the Bavarian government nor the newly formed federal government were willing to lend support to the project. Loeffelholz had asked for financial support from many different departments and had even managed to get through to the offices of the recently elected Federal Chancellor Konrad Adenauer, but he ultimately failed in his mission. Despairingly, Loeffelholz wrote an open letter to the Atlantropa Institute and the chancellery that he would resign from his office, as he saw no potential for success. He also emphasized that he had concluded that Atlantropa lacked a sufficiently thorough scientific basis. This public disavowal of the project by one of its former supporters was a devastating blow to Sörgel and his institute.[47]

One year later, in December 1952, the creator of Atlantropa died in a bicycle accident. His institute continued to exist for a little while longer, but nobody could—or was willing to—take over the reins and keep the project alive. In 1960, the designated trustee of the institute, Hans Aburi, liquidated the few remaining assets and published a eulogy to Atlantropa with the title "The End of a Big Idea." While it had once promised so much, Aburi wrote, Sörgel's project now seemed outdated in the age of nuclear energy. The name Atlantropa was once again used in 1963 by an obscure and short-lived magazine on Euro-African unity and cooperation that had very little in common with Sörgel's project other than the belief that the Sahara could be cultivated. Sörgel's dream was quickly forgotten in a new German republic that was embarking on swift economic growth after the devastating war and whose eventual participation in European cooperation would look quite different from what Sörgel had imagined.[48]

Civilization and Climate

Sörgel's project had been a product of its own specific place and time— Germany of the late 1920s and early 1930s, where the crises of the first half of the twentieth century were most deeply felt and where extreme solutions were sought with a kind of millenarian enthusiasm. Though Atlantropa's vision for a hydropower-fueled new order faded, the project endured in other ways. While most of Sörgel's ideas remained remarkably stable and unchanged over the years, he did put increasing emphasis on aspects of climate change—both "natural" and engineered. When writing about his project after the Second World War, Sörgel emphasized climate modification as a force to pacify the world: "One day it will be incomprehensible how impotent we once were . . . when the climate of the earth will have been improved by surprisingly simple and elementary means and thus the overexcited nerves of all against all will be calmed."[49]

Sörgel had been interested in this aspect of Atlantropa from the earliest days. He had put his own project in line with Roudaire's *mer intérieure* and had adopted the French engineer's arguments on both the alleged desiccation of the Sahara and the possible restoration of a more moderate climate in the desert. During the 1930s, Sörgel had deepened his knowledge on climate change and climate modification. He collected articles and, among other topics, read up on carbon dioxide's role in regulating the temperature—an idea that was only just coming back into fashion after Svante Arrhenius's findings had been largely forgotten, ignored, or dismissed for some decades. Sörgel wrote ample notes on climatic conditions in Africa, highlighting the growth of the Sahara. Desiccation, as Sörgel cited from a popular book on Africa, was a natural process that had been "sped up enormously by the lack of judgment and the greed of humans." Sörgel also familiarized himself with current ideas of climate modification that Nazi planners tried out in both Germany and the occupied areas in the East.[50]

After the war, Sörgel incorporated this information into his Atlantropa propaganda. The main task of the project remained accommodating European settlers in Africa, but now the emphasis shifted slightly from desert irrigation to the "improvement of the climate" as the most important precondition for this colonial endeavor. Through his collaborator Theens, Sörgel got hold of

Anton Metternich's *The Desert Threatens*, an apocalyptic warning of global desiccation and food shortage. The book, according to Sörgel's reading, was really "nothing other than a call for Atlantropa," and there were indeed common themes in the diagnosis of a looming disaster for occidental culture and the calls for quick and concerted action to avert catastrophe. Only radical active geopolitics and the transformation of large-scale climatic conditions, Sörgel argued, could avert the realization of Metternich's fears about the advancing desert.[51]

The issue of climate engineering also led to the most serious debate on the Atlantropa project in the postwar years. Sörgel's attempt to explain the climatic changes that the engineered African inland seas would produce fell short of scientific standards. An article in one of the leading geographical journals attacked the vagueness of Sörgel's calculations in general, criticizing the lack of climatological and hydrographic material that would support the viability of the project. This exposed the major weakness of Atlantropa that Sörgel never managed to overcome: the lack of scientific expertise and data. In fact, Sörgel was very interested in climatic theories, but he remained rather oblivious to the details of climatology. When the authors of the journal article charged him with using isothermal maps of Africa to give a false appearance of scientific depth, they had struck Sörgel's Achilles' heel. The allegations of shoddy work did not stop there: the geographers Troll, van Eimern, and Daume added that Sörgel had vastly overestimated the cultivable area in northern Africa and exaggerated the climatic changes that Atlantropa would engender. Van Eimern compared the enlarged Lake Chad with the Red Sea, whose climatic effect does not reach far beyond its shores. And Daume criticized Sörgel for using outdated data to measure the time it would take to lower the Mediterranean.[52]

Probably most damaging to Sörgel's aspirations was a contribution by Hermann Flohn, who went on to become one of the leading German climatologists after the war and had already published on anthropogenic climate change and climate modification during the Third Reich. In 1951, Flohn challenged Sörgel's claims that standing water in the dry belt of the earth would bring about any noticeable climate change. Moreover, he argued, any human attempts to change climatic conditions always had unforeseen detrimental effects, like the lowering of water tables or increased soil erosion—even from local interventions: "Should not the engineer, in particular," Flohn asked, "draw some humility or even awe of natural forces from these small-scale processes when deliberating on the large scale?"[53]

Humility had never been Sörgel's strong suit, but Flohn's attack was still not entirely fair. Sörgel had, in fact, considered the detrimental effects of human intervention. In the 1930s, he closely followed the debates on *Versteppung,* or desertification, that the "landscape advocate" Alwin Seifert had initiated. Seifert's main argument was that human water engineering schemes had disturbed the natural equilibrium. The unrestrained use of technology, he contended, had led to a progressive desiccation of Germany, which manifested in decreasing water tables and the slow but continuous change to a steppe climate. Sörgel incorporated Seifert's terminology into his own publications. In *Die drei großen "A,"* he justified redirecting the Gulf Stream and thus producing a warmer and more humid climate with the claim that Europe was suffering from *Versteppung*. Sörgel even wrote an article on the phenomenon in 1937. "Blood and Soil," as it was called, contained a mix of Sörgel's approximation of Nazi vocabulary and a rather idiosyncratic reading of Seifert's essay. The architect avowed that desiccation was ongoing but added that it moved from east to west—a notion that was widespread during the Nazi period but was in fact not one of Seifert's original claims. Sörgel ended his essay with yet another call for the application of large-scale technology, which alone could fight desertification and save Europe.[54]

In the end, Sörgel's project remained unrealized: the Gibraltar dam was never built, the Congo River did not become a lake, and Europe and Africa never merged into Atlantropa. It might be tempting to simply dismiss the whole episode as a bizarre and ultimately inconsequential footnote to the history of the twentieth century. That, however, would not do Atlantropa justice. Sörgel's views about the primacy of technology, his idea of active geopolitics, and his enthusiasm for environmental and cultural transformation on a gigantic scale undoubtedly struck a chord in the 1920s and 1930s. While commentators criticized the project as unrealistic on political grounds, it was more than just the far-fetched colonial dream of an overly ambitious German architect with a god complex. Atlantropa became the subject of a lively, if short-lived, debate in the news of the time. And while many commentators regarded the project as a whole as too big to be realized, its constituent parts enjoyed a more successful and long-lasting history. As inflated as the scale of Atlantropa was, it reflected—and refracted—many of the public and academic discussions of political, social, and environmental conditions. In both trying to solve the pressing issues of the time and attempting to convince as many Europeans as possible of his vision, Sörgel incorporated the most current and fashionable strains of thought into his Atlantropa propaganda, among them a blind faith in technology and a stalwartly racist colonial vision. Once completed, the project would allegedly solve the issues of parliamentary stalemates, nation-state rivalries, geopolitical imbalances, European overpopulation, and environmental degradation in one fell swoop.

Atlantropa also exemplifies the fate of the climate change debate that had seemingly run out of steam in the early years of the twentieth century, only to quickly reemerge in altered form in the 1920s and 1930s. While the academic discussion of large-scale climatic shifts and desertification took some more time to recover from the temporary consensus that there was neither hard evidence nor a good causal model for a global, unidirectional process of desiccation, the terms and concepts of the debate had found their way into the popular imagination. When Sörgel wrote about the once fertile swaths of North Africa, he referred to a well-known idea that needed no elaboration for his audience. For many Europeans, the view that the earth was mutable and that environments could drastically change had become conventional. Sörgel used climate change not simply to explain an environmental degradation that had taken place but also to justify his own project of colonial climate engineering. If macro-climatic conditions could change by either natural or anthropogenic causes, what spoke against using modern technology to alter the climate in a desired direction? While Sörgel took this argument to an extreme in Atlantropa, he reflected a more widespread interest in climate engineering. Even according to Atlantropa's critics like Kurt Hiehle, the "climatization" of arid landscapes appeared to be the most important task, and the "Faustian struggle against the desert" the most worthwhile endeavor of humankind in the second half of the twentieth century.[55]

As a self-proclaimed proponent of this allegedly lofty aim of environmental engineering, Sörgel was one of the ideological forefathers of today's geoengineering advocates, both in putting large-scale technology on a pedestal and in giving short shrift to its intended and unintended consequences on the ground. And once again Sörgel was not alone. The same popular debates on climate change, desertification, and *Versteppung* that Sörgel used in his propaganda also inspired planners in the Third Reich to think about colonial environmental transformation on an unprecedented scale. While Sörgel always focused his energies on Africa, however, Nazi bureaucrats saw Eastern Europe as the proper space for their designs to "Germanize" landscapes. Their big moment came with the German occupation of the *Ostland*, the land of the East.

6

Slavic Steppes and German Gardens

DESERTIFICATION IN THE THIRD REICH

"I FIND TRUE heroism in my father's attempt to make this desert habitable and blessed," Frederick the Great once said about his predecessor's project to settle the easternmost parts of the Prussian kingdom. In fact, Frederick was so inspired by his father that he continued and expanded the efforts. About 150 years later—in 1920—the young German landscape architect Heinrich Wiepking-Jürgensmann sang the praises of the Prussian king, whom he declared to be "the greatest German colonizer." At first, Wiepking's paean may seem neither very surprising nor noteworthy: after the double shock of the lost world war and the harsh terms of the Versailles Treaty, and in a period of political unrest and economic instability, many Germans looked to the past for inspiration. Frederick the Great, the enlightened monarch of the eighteenth century, may seem like a comparatively innocuous choice in the context of the radical politics of the Weimar period.[1] And yet a closer look casts Wiepking's statement in a rather different light. In his short article, the landscape architect chose not to imagine Frederick the Great at a courtyard table in his leafy summer residence at Rheinsberg, conversing with his friend and correspondent Voltaire—an image that has become almost iconic in the popular rediscovery of the Prussian past in Germany over the past thirty years. Instead, Wiepking foreshadowed the later Nazi appropriation of Frederick the Great by putting the king in a line of German colonizers in the East, referring forebodingly to his "German blood." A few pages further on, Wiepking connected past and present explicitly, drawing the allegedly self-evident conclusion that new German settlements in the East were essential to fortify the country "against Sarmatian and Hunnish barbarism [*Unkultur*]."[2]

This characterization of Frederick the Great grows even more ominous in view of Wiepking's biography. Having abandoned his architecture studies in Hannover after just one year in 1912, he first interned and then worked at a famous landscaping firm. In 1922, Wiepking moved to Berlin, where he worked as a self-appointed architect until his big moment arrived in 1934: through connections in the bureaucracy of higher education, he obtained a professorship in garden and landscape architecture at the Agricultural University of Berlin. In 1941, Heinrich Himmler called upon Wiepking to become a "special deputy"—a high-ranking post at Himmler's Reichskommissariat für die Festigung des deutschen Volkstums (RKF), whose name translates somewhat awkwardly as the Reich Commissariat for the Strengthening of Germandom.[3]

With his immediate superior Konrad Meyer and his colleagues at the RKF, Wiepking was responsible for commissioning, drafting, and collecting plans to completely reorganize and transform the Nazi-occupied areas in the East—a task he took on with enthusiasm. In one of his articles for the SS magazine *Das Schwarze Korps*, he described the problem that needed to be tackled: "The specter of the steppe [*Steppengeist*], the rapacity, has progressed far into the land! It is imperative for the sake of the *Volk* to counter that specter with all means available." This statement, conflating environmental and cultural decline, bears some similarity to Sörgel's worries about comprehensive European decline. And like Sörgel, Wiepking was fond of playing on Spengler's rhetoric in predicting the "destruction of the occident." Both Sörgel and Wiepking also referred to the controversial theories about desertification, or "steppification," by their contemporary Alwin Seifert; and both looked for colonizable and modifiable land beyond Germany's political borders—the former to the south and the latter to the east. Very much unlike Sörgel, however, Wiepking managed to reach a position in the Nazi administration that brought him close to the levers of power. When the German army occupied Poland and then pushed on further toward the east, Wiepking was finally in a position to not only *portray* a great colonizer of the eastern steppe as he had done in 1920 but to become one himself.[4]

Rather than looking at the origins of RKF policies within an elusive canon of National Socialist ideology or uncovering the bureaucratic mechanisms behind the decision-making processes, into the following two chapters I put the Nazis' plans to totally reorganize the East into the context of ideas about environmental change and their link to ideas about cultural development. As Timothy Snyder has argued, the Nazi project of genocide and expansion was at least in part a reaction to a felt ecological crisis. Wiepking's biography, his writings, and his actions as a member of the RKF are all important to understand where that sense of crisis came from and what shape it took from the 1920s to the 1940s.[5]

The German ideologues and planners in the Third Reich borrowed and developed concepts from geology and climatology in addition to the life sciences. They were, after all, concerned with both "blood" *and* "soil," which planners connected through *Versteppung*, the progressive deterioration of the environment through the advance of steppes and steppe climates. While Nazi understandings of climate change had come a long way from the nineteenth-century debates over a potential desiccation of North Africa, the two historical moments were connected by concepts, anxieties, and even individuals. Ideas about *Versteppung* grew—at least in part—out of the nineteenth-century climate change debate. Yet they appeared in Nazi Germany in a new, radically racialized and militarized guise.[6]

To be sure, hypotheses on climate change and climate engineering had never belonged to a sphere of pure science and engineering—an imaginary space that never existed outside the realm of rhetoric in any case. In fact, as I have shown in the preceding chapters, climate and climate change had always been deeply political; ideas about desiccation originated in colonial environments, and colonial engineers intended to transform those same environments to make room for European settlement and societal transformation. In the Third Reich, however, the political charge of changing climates reached unprecedented dimensions. In their documents, RKF planners combined active climatic modification with the political project of military occupation, forced resettlement, and genocide.

To assess Jürgen Zimmerer's claim that the Nazi *Ostland* was born "out of the spirit of colonialism," it is important to focus on the practices and approaches of colonial science, technology, and planning that were deeply intertwined with violence. Climate change, desertification, and large-scale environmental transformation—all shaped in colonial or proto-colonial contexts in nineteenth-century debates—became key concepts in Nazi theories and policies of colonization and their legitimization. In the process, the concept of "desertification" or *Versteppung* itself changed. Now fully removed from its academic home in geology departments, it was absorbed into the practical and application-oriented realm of landscape architects and "nature advocates"—the latter one of the many distinctive professions created by the Third Reich. Two of the most outspoken members of this new elite of landscape experts were Heinrich Wiepking and his rival Alwin Seifert, who rose to fame in the 1930s with his landscaping work on the new Autobahn, the German highway system. Despite their personal differences, Seifert and Wiepking shared a strong interest in *Versteppung*, which for them became the ultimate sign of a maltreated, exploited, and—especially in Wiepking's writings—un-German landscape that had to be saved through large-scale planning: in Wiepking's version of Nazi ideology, "unworthy life" and *Versteppung* represented the antitheses to valuable Germanic "blood and soil." And planners had to turn these unhealthy deserts and steppes into healthy Germanic forests and fields to allow Germans to prosper in the East.[7]

The Mythical "German East"

The German interest in *Versteppung* and landscape planning in the East did not arise in a vacuum. It resulted from the convergence of different strands of thinking with their own particular histories. Conservative and *völkisch* thinkers combined their growing obsession with the German *Drang nach Osten* (the "drive eastwards")[8] with a widespread and long-standing belief in Germans' special relationship to nature, for which in turn they drew on ideas from an eclectic range of nature conservation, or *Heimat*, associations. This combination already provided a pronounced environmental dimension to German claims to the East, which often included references to a particular Germanic bond with nature. In the 1920, some *völkisch* writers further supported these claims with theories and anxieties about desiccating soils and deteriorating climatic conditions threatening both German agriculture and culture from the east. At the same time, right-wing revisionists elevated the German *Drang nach Osten* from a political prerogative to an environmental—and even a cultural—necessity to ensure German survival.[9]

The "drive eastward," that is, the idea of an inborn urge to seek new colonial living space in the East, was of course not itself ahistorical. Rather, it was the most prominent and dynamic part of Germans' growing fixation on Eastern Europe and Russia in the first half of the twentieth century. It exhibited a rather paradoxical mix of "irrational yearning and anxiety," which painted lands in the East as simultaneously a potential paradise and the origin of all evil. Geographers played an important role in propagating these images. And as a subfield of geography, climate science had already become a vital part of the German debate on the East around the turn of the twentieth century. In 1904, Joseph Partsch, known for his geographical and climatological work on the Mediterranean region, had published a book on "Middle Europe." Partsch had already argued for regional, anthropogenic causes for environmental changes in North Africa in earlier work and now ascribed the alleged differences between German and Slavic landscapes to the cultural standing of their inhabitants. But while the Germans had been "superior in every area of work" throughout history, he contended, now the East loomed as a growing and existential threat to Middle Europe.[10]

Partsch did not construct his anxieties out of thin air, as the preoccupation with lands to the east was not wholly new in the early twentieth century. He, along with others, built upon earlier Enlightenment ideas of a cultural border between Germans and Slavs, which itself reflected even earlier notions of cultural difference. The fixation on the East became more firmly grounded in German cultural discourses in the nineteenth century and then gathered even more momentum—and a new aggressive overtone—in the wake of early

German advances on the eastern front during World War I after the myth-laden Battle of Tannenberg. Germany's temporary occupation of vast tracts of land spurred fantasies of an expanded Greater Germany that would stretch to the outer reaches of Europe in the East. From the start, these fantasies were marked by the impulse to impose order on landscapes that the authors perceived to be unordered, deficient, or even morally depleted.[11]

The German soldiers, as Vejas Liulevicius traces in eyewitness accounts, were surprised about what they regarded as the backwardness and uncultivated nature of the eastern landscapes. Governmental and lay planners saw the opportunity to reorganize the eastern lands according to their own principles. The early schemes of a new, reordered East already showed a striking disrespect for the resident populations and distinct authoritarian inclinations, reflecting the colonizing tendency to view spaces as empty and completely malleable. But the planners quickly ran out of time: when Germany surrendered in 1918 and then signed the Treaty of Versailles a year later, their designs became obsolete in an instant.[12]

And yet something of the "spirit of Tannenberg" survived into the postwar years, and eventually became stronger than ever. The enthusiasm that planners felt after the German invasion of Russian soil in 1914–15 quickly evaporated, but their disappointment over both the eventual outcome of the war and the political pragmatism of the Weimar governments spawned a renewed interest in the East. Spurred on by both the "unjust" loss of German territory and stories of Polish atrocities against Germans in some of the transferred territories, conservative and *völkisch* authors published tracts on German history and culture in the East in which they told of the past heroic greatness of Germany and of the possibility of a rebirth of that greatness after the ignominious defeat in the First World War. The writings ranged from cheap dime novels to serious academic treatises, but the argument—whether explicitly stated or not—was similar across much of the literature: Germany, now smaller, left to its own devices, and encircled by hostile states, needed to achieve economic autarky, which called above all else for a self-sufficient food supply. This was possible only through obtaining more agricultural land. After the definitive loss of the German overseas colonies in the Treaty of Versailles, the colonial rhetoric now focused on the plains of the East as a new target. And these plains, a fast-growing body of literature argued, had in any case been German since the times of the Teutonic Knights in the thirteenth century.[13]

In addition to the constantly repeated story of the supposed heroism and moral fortitude of the Teutonic Knights, the authors laid claim to the *Ostland* through various rationalizations. One of the most persistent was an environmental argument that would later become very important for Heinrich Wiepking and his colleagues in the RKF. In order to assert a moral right to the East,

German authors often described the squalor that the land had allegedly fallen into once the order of the Teutonic Knights—and with it German influence—went into decline in the fifteenth century. The conservative historian Karl Josef Kaufmann argued in 1926 that under the Poles and other Slavic peoples who had taken over, the "fields grew over with weeds, the animals degenerated, and the people lived lives that were not much better than those of their animals with whom they shared their degraded huts." Here, Kaufmann built upon arguments by his famous nineteenth-century predecessor Heinrich von Treitschke, who had described Slavic rule over Germans as an "unnatural state of affairs," referring to the "desolating rule of the Poles." Kaufmann further expounded Treitschke's evaluation of the state of the land by contending that the soil had been converted into a desert wherever Poles had taken power.[14]

The moral of this narrative was difficult to miss: once again, non-Germanic peoples, who did not know how to care for land or forests, threatened the landscapes Germans had lost in the Treaty of Versailles. While dense forests had been the prime obstacle for German settlers in the East during the twelfth and thirteenth centuries, in the twentieth century *völkisch* writers now regarded steppes and deserts as a kind of frontier to be ordered, civilized, and transformed. Without Germans to rule and control the landscapes, these writers argued, the valuable territory would deteriorate into wilderness, lose its fertility, and become a barbaric desert. Kaufmann ended his chapter with the following, programmatic words: "Which *Volk* has the rightful claim to a land; the one that has brought culture and bloom to the land, or the one that has thrown it into ruin and decline, has turned it into a desert?" Thus, Kaufmann argued, Germans had not only a historical-racial claim to the land but also a cultural-environmental one.[15]

German Nature, Cultural Soil

Kaufmann's argument that Germans had a right to land in the East was predicated on the belief that Germans had a special relationship to nature, that there was something in their Germanic heritage or their blood that made them understand nature more fully than other peoples or races. Kaufmann and others developed this concept from early nineteenth-century ideas of distinctive national "natures" or environmental features proclaimed by such luminaries as Johann Gottfried von Herder and Alexander von Humboldt. Nationally minded Germans clung to these ideas as compensation for the lack of monuments from early Germanic cultures, which had not produced anything as impressive as Greek temples or Egyptian pyramids. In their rhetoric, they inflated the simple and quite plain megalithic tombs of old Germanic tribes into great achievements of German culture and put them on a par with, if not above, the Acropolis and the Great Pyramid of Giza. The "natural" look of the boulder graves, which blended seamlessly into their environment, became evidence of German cultural superiority, which aimed not to consciously impose culture onto nature but rather at an "adaptation of culture to nature," as Joachim Wolschke-Bulmahn has argued.[16]

During the Weimar years, researchers and writers often contrasted the portrayal of "desolate" eastern landscapes with the proposition of a special German bond with nature. As they depicted the East ready for environmental transformation and colonization, the region took on more and more of a mystique of the frontier. And like its famous counterpart in the United States, the frontier in the East offered, in addition to wild and unordered landscapes, spaces free of the purported ills of civilization. The German would-be colonizers thus created an ideal imaginary territory, which was both deficient and receptive to transformation. Meanwhile, Germans also amassed practical experience with land reclamation projects during the Weimar Republic. Rather than focusing on eastern deserts or steppes, the efforts were mainly directed toward moors and peat bogs within the political borders of Germany. The Weimar governments continued efforts from the eighteenth and nineteenth centuries to drain waterlogged land, especially in northern Germany. Aside from making deposits of peat accessible for harvesting, this was also an attempt to fill the supposed need for more cultivable land, which—already in 1927—was sometimes elevated to a question of survival. Despite the rhetoric, however, the efforts did not achieve their goals. According to data collected by the government, the area of uncultivated land in Germany actually increased over the period of the Weimar Republic.[17]

Despite the lack of success of these reclamation projects during the Weimar period, they anticipated the imminent future. During the First World War, the German government had forced prisoners of war to perform the manual labor in the marshes. At one point more than fifty thousand forced laborers were working in reclamation projects. After the war the government maintained the practice, but the laborers were now civilian prisoners instead of prisoners of war. Hitler's government continued both the attempts at land reclamation in marshlands and the use of forced labor. Another important corollary of the Weimar reclamation attempts was a debate about environmental consequences of draining water from marshlands, which would ultimately contribute to the controversial debate on *Versteppung* in the 1930s.[18]

In the interwar years, aggregated ideas about desertified landscapes, the special relationship between Germans and nature, and the long history of German culture in the East found an institutional home in the Leipzig-based Stiftung für deutsche Volks- und Kulturbodenforschung (SdVK), or the Fund for German National and Cultural Soil Research. Among its founders were Joseph

Partsch and his friend and colleague Albrecht Penck, the famous German geographer and climatologist. Both men had played prominent roles in the climate change debate around the turn of the twentieth century. In 1926, Penck, by this time a highly decorated professor in Berlin, assumed the first presidency of the SdVK. A student of the geologist and desert explorer Karl Alfred von Zittel, he had become a household name in glaciology after his collaborative work on the succession of ice ages with his own former student, the climatologist Eduard Brückner. Penck supported Brückner's work and his findings of periodic variations in climate, but—like Partsch—was highly critical of the sensationalist theories of a progressive, worldwide desiccation that the American geographer Ellsworth Huntington publicized in the first quarter of the twentieth century. Penck also did not vocally endorse the theories of environmental decline and *Versteppung* in Eastern Europe, which some of his contemporaries started to advance (nor did Penck discourage their work).[19]

And yet Penck played a central role in developing conceptions about the "German East" that Nazi landscape architects and planners like Wiepking would later elaborate. Encouraged by his teacher Zittel, Penck had shown an early interest in geography and had proven his *völkisch* potential and his interest in questions about the *Ostland* with an unmistakably political article on "German *Volk* and German Soil" in 1907. After the First World War, he had become interested in Ratzel's anthropogeography, and especially his ideas about *Lebensraum*. One of Penck's articles building on Ratzel's approach appeared in a collected volume, edited by none other than the desert geomorphologist Johannes Walther, who had become the director of the German society for natural scientists, the Leopoldina. Reflecting the national discontent about the losses of territory in the Treaty of Versailles, Penck had been turning toward a more political dimension of geography over the 1920s. This political stance was also an explicit part of the SdVK, despite the rather implausible public assertions that its work was "purely scientific" and "free of any bias and any political approach." Elaborating on Ratzel's view of open spaces as "life-sustaining," Penck supported German territorial claims, arguing that "a growing *Volk* needs space"—space that was to be found in the East.[20]

Penck did not simply adapt Ratzel's theories to the contemporary political situation. With his authority as an internationally respected scholar, Penck also contributed his own ideas—or at least his own catchphrases—to the political debate. Borrowing from botany and agronomy and the work of Brückner, he brought together the terms *Volksboden* and *Kulturboden*—or "national soil" and "cultural soil"—that would become key concepts for *völkisch* ideologists. While the *Volksboden*, demarcating the extent of the German language being used by the local population, could change with political circumstances, the *Kulturboden*, demarcating the extent of German culture imprinted on the land, was

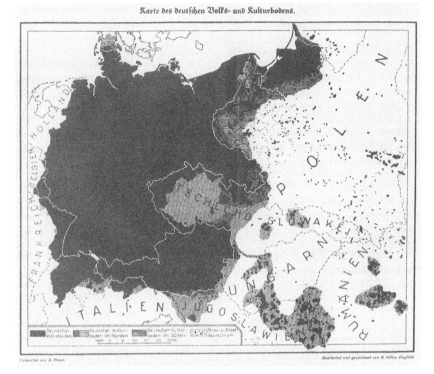

FIGURE 6.1. Penck's map of the German Volks- und Kulturboden, colored in black and gray, respectively. *Source*: Albrecht Penck, "Deutscher Volks- und Kulturboden," in *Volk unter Völkern: Bücher des Deutschtums*, ed. K. C. Loesch (Breslau: Hirt, 1925), 72.

more constant. Penck argued that the *Kulturboden* exceeded the extent of the *Volksboden* almost everywhere in the East, providing evidence of former German presence and cultural activity. Penck illustrated this with a map showing the extent of German "cultural soil" reaching far into the East and including, among other territories, the entirety of Bohemia and Moravia (Figure 6.1).[21]

Building upon claims of the special German relationship with nature, Penck argued that each *Volk* had its own, unique relation to the soil: thus, the "cultural soil" could give clear evidence of who had been responsible for its improvement or deterioration. This amounted to an implicit claim to the "German soil" in the East only thinly veiled as a general theory. In the East, Penck argued, the soil was well taken care of as far as German settlements extended. The difference between German and Slavic landscapes was allegedly clearly visible even from the window of a train. Stylizing the German landscapes as both beautiful and functional, Penck rhapsodized the German "cultural soil" as the greatest accomplishment of the German people.[22]

The SdVK and other *Ostforscher* expanded their institutional reach and created research centers during the Weimar Republic. While not all of these institutions had expansionist or revisionist agendas, they did contribute to an organizational structure that the Nazis and their academic collaborators would later use. Penck himself continued his work on German "cultural soil" in the East, which culminated in the mammoth project of the Statistical Atlas of the German People in Middle Europe. Penck and his colleagues in the SdVK and the Prussian Academy of Sciences started the project in 1930 and continued their work in the Third Reich under the new title *Atlas of German Living Space in Middle Europe*. The collection of maps, showing the alleged extent of German "cultural soil" in the East, was finally published in 1937, in time to serve the preparations for war that Nazi Germany had embarked on. While Penck did not have a role in the Nazi bureaucracy and did not endorse the use of his work for any military purposes, he—among other members of the Prussian Academy—was quickly ready to serve Nazi interests with his project of the atlas and with other publications warning of a limited carrying capacity of the world, which made German imperial expansion an even more pressing issue.[23]

Blood and Soil

Hitler's accession to power provided a new field of opportunity for the *Ostforscher*. Scholars working on questions of the "German East," like the historian Hermann Aubin, now found a conducive atmosphere for their revisionist worldviews. Institutions like the *Bund Deutscher Osten*—the League of the German East—and the *Ostuniversitäten*—the Universities of the East—flourished in the 1930s. During the first few years of the Third Reich, *Ostforscher* shed their pretense of objectivity and became a self-consciously political force that contributed ideological backing and blueprints for Nazi *Ostpolitik*, the program of expansionism into a hazily defined and delimited "East." Through this new dynamism of research on the East, the German focus on eastern soils, already a prevalent theme during the Weimar years, became even more prominent.[24]

But just as the National Socialists did not have a clear idea of the extent and content of *Ostpolitik*, the new regime's emphasis on both blood *and* soil also created an uneasy tension. With this dual focus, the German *Volk* was either an immutable race of the same timeless blood or a geographically determined race dependent on the characteristics and potential changes of the soil and other environmental conditions. Richard Walther Darré, in particular, attempted to combine these two potentially contradictory parts of Nazi ideology.[25] As the first minister of food and agriculture in the Third Reich, Darré became the main proponent of "blood and soil" ideology, or the belief that both blood, or race, and the soil, or the particular characteristics of landscapes, determined the

welfare and the well-being of a *Volk*. Darré had coined—though never conclusively defined—the phrase in his 1930s book *New Nobility of Blood and Soil*,
which would become a reference work for Nazis who developed policies on
agricultural and landscape planning. Alongside his eugenicist ideas about controlled and regulated human breeding, Darré asserted that Germans had a special talent to care for nature, a claim he mixed with his own brand of anticapital-
ism. In 1930, this blend of neoconservatism, racism, and pseudo-socialism
remained in the realm of theory. As a Nazi minister and the creator and head of
the Race and Resettlement Office in the Third Reich, however, Darré was in a
position to exert a strong influence within the Nazi bureaucracy. His suggestions of a new model of *völkisch* farming predicated on both racism and anti-
capitalism would become an important point of reference for the Nazi officials
who planned the large-scale resettlements and the environmental reorganization of the East.[26]

In the end, Darré himself did not rise to become the most important figure
behind the German planning efforts in the annexed and occupied territories.
In the ubiquitous struggles between different departments and offices of the
Nazi bureaucracy, Himmler outmaneuvered his competitors and gained control over large parts of the German plans to reorganize the East. In 1937, he had
already overseen the coordination and consolidation of the various institutes
and offices dealing with *Ostforschung* in the Ethnic German Liaison Office (the
Volksdeutsche Mittelstelle), headed by the SS officer Werner Lorenz. In 1939,
only five weeks after Germany invaded Poland, Himmler added the title of
Reich Commissioner for the Strengthening of Germandom (RKF) to the already long list of positions he was holding. In this role he created an administration that would eventually oversee the reorganization of the eastern territories that Germany began to invade and occupy that same year.

The RKF's Main Department for Planning and Soil was led by the agronomist
Konrad Meyer, who cooperated closely with his "special deputy," the landscape
architect Heinrich Wiepking.[27] Meyer, a professor at the universities of Jena and
Berlin, rose swiftly through the Nazi bureaucracy to become one of the most
senior planners of the SS. Under his guidance *Landschaftspflege*, or "care for the
landscape," became the central planning discipline to which all other departments were subordinated. Meyer's planners internalized the emphasis on both
"blood and environment [*Umwelt*]," as Wiepking put it. Meyer himself also contributed to the emphasis on soil, landscape, and climate in the Nazi evaluation
of the East. A 1941 volume edited by Meyer showed an aerial photograph of what
he described as German and Polish areas of settlement (Figure 6.2), the former
marked by orderly fields and large wooded areas and described as "shaped in
accordance with the natural circumstances." The Polish area, on the other hand,
Meyer depicted as carved up into "arbitrary" and "unseemly" parcels. Similar

Diesseits, südlich der ehemaligen Grenze ist die Landschaft gestaltet nach den natürlichen Erfordernissen und nach dem kolonisatorischen Willen und Vermögen der einzelnen Gemeinden, größen- und lagemäßig einwandfrei, wenn auch im Zuge einer neuzeitlichen Planung verbesserungsfähig — jenseits der ehemaligen Grenze breitet sich eine willkürlich und schematisch bis in kleinste, unansehnliche und unwirtschaftliche Stücke zerschnittene Fläche aus, besetzt mit baulich völlig zersplitterten, regellos angefüllten Dörfern.

Freigegeben durch RLM 1761/40

FIGURE 6.2. "German and Polish Settlement Areas." Aerial photograph with the German area in the lower and the Polish area in the upper part of the image, 1941. *Source*: Konrad Meyer, ed., *Landvolk im Werden: Material zum ländlichen Aufbau in den neuen Ostgebieten und zur Gestaltung des dörflichen Lebens* (Berlin: Deutsche Landbuchhandlung, 1941), 272.

Zwei Luftaufnahmengeben Ein-
blick in die bäuerliche Landschaft des Warthegaues. Frucht-
bare Felder wechseln mit Wäl-
dern und Seen ab. Es handelt
sich hier um eine ausgespro-
chen deutsche Kulturlandschaft.
Luftaufnahmen der östlich der
Reichsgrenze vor 1918 gelege-
nen, vornehmlich polnischen
Gebiete zeigen deutlich die un-
regelmäßige Anlage der Felder
und die schlechten Straßenver-
hältnisse. Schon im Grundriß
der Landschaftsgestaltung zeich-
nen sich die beiden Welten von-
einander ab: Deutsche Ordnung
und „polnische Wirtschaft".

FIGURE 6.3. "German [top image] and Polish Settlement Patterns," 1941. *Source:* Fritz
Wächtler, ed., *Reichsaufbau im Osten* (Munich: Deutscher Volksverlag, 1941), 166.

depictions became staple images in Nazi writers' efforts to legitimize large-scale
planning in the East (Figure 6.3).[28]

The images did not always show clearly what exactly was wrong in the territories under Polish and Slavic influence. Aerial photographs and drawings
portrayed the alleged lack of a well-defined structure of Polish settlements,
farms, and fields but did not give much information about the environmental
conditions on the ground. The accompanying texts, however, made the

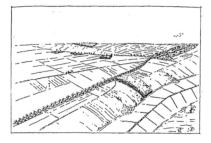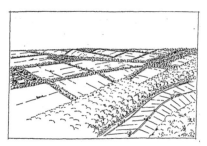

FIGURE 6.4. "Unformed Cultural Steppe" and "Shaped German Cultural Landscape," 1941.
Source: Werner Junge, "Aufbauelemente einer deutschen Heimatlandschaft," in *Landvolk im Werden: Material zum ländlichen Aufbau in den neuen Ostgebieten und zur Gestaltung des dörflichen Lebens*, ed. Konrad Meyer (Berlin: Deutsche Landbuchhandlung, 1941), 106–7.

author's position abundantly clear: Slavic rule had turned most of the East into an unproductive and infertile wilderness. This wilderness was sometimes described as waterlogged moors or marshes, which Nazis claimed was an indication of Slavic laziness or inability to install drainage systems and undertake large agricultural projects. Some articles even argued that "swamp formation [*Versumpfung*]" was the main threat in the East, caused by irresponsible deforestations. The destruction of forests, however, remained a prime focus in the search for causes of desiccation.[29]

Nazi planners apparently did not notice—or at least did not openly acknowledge—any inconsistencies in their arguments and could refer to literature from the 1920s to support their claims. Sokolowski, who supplied both Sörgel and Darré with images of an impending environmental catastrophe, had written in 1929 that the destruction of agricultural land in the East was due to "deforestation, desertification, [and] swamp formation." Nazi descriptions reflected this analysis, describing the eastern landscape as either too wet or too dry, but in any case deficient. In the RKF, however, one particular deficiency of the landscape became the focus of concern above all others.[30]

Meyer's volume on landscape planning included an article by Werner Junge, who discussed what made a landscape "German" and illustrated this with two images: one of the "shaped German cultural landscape of the future," featuring parcels of forests and neatly structured rows of trees and hedges, and the other one of the "unformed cultural *steppe*" before the arrival of the Germans—a denuded landscape lacking both order and trees (Figure 6.4). As an example of how the latter landscapes could become more like the former, Junge cited Alwin Seifert's work in the construction of the German Autobahn as an exemplary case of "creative landscape design." It was Seifert, as well, who had popularized the terms *Steppe* and *Versteppung* that Junge used prominently in his essay. These terms would become central to the RKF's plans for a trans-formation and "Germanization" of the East.[31]

Versteppung

When the German army invaded Poland in 1939, it quickly became clear that no ordinary military occupation would follow. Over two decades, *Ostforscher* and *völkisch* German authors had prepared the ground for this moment. De-mands from Nazi officers for German colonization and a radical transforma-tion of the East followed on the heels of the German invaders across the Polish border. And even the German foot soldiers themselves were now equipped with a standard vocabulary that they could use to confirm their preconceived notions—or as one 1940 article on the "design tasks in the new East Prussia" put it:

> When we . . . crossed the Polish border from East Prussia . . . , we believed we had arrived on a different continent, and it was not so much the miser-able cottages and huts and the filth in the towns made up of iron sheet-covered shacks that evoked this feeling, but rather the complete bleakness of this *barely cultivated steppe*, this landscape entirely stripped of any beauty. Even the simplest man felt that here *an ordering hand, a planning and creative will, has always been lacking.* . . . The Pole is, after all, condemned to artistic infertility, because he only misuses nature and has never strived responsibly toward the inner law of life.[32]

Germany sought military and political control in the East, but also control of the environment, the transformation of the steppes. During the first years of the Third Reich, writers of publications on the East had increasingly expressed the difference between German and non-German landscapes as a function of the quality of their soils and the corresponding landscapes. Under Slavic influence—the common argument went—the East had become desertified or, more to the point, "steppified."[33]

The steppe, as a descriptive term for the landscapes in the East, had gained prominence in Germany in the 1930s. But despite the increasingly common usage, the term itself remained ill defined. The term of murky origins describ-ing the largely treeless plains in Central Asia had been popularized in German by Alexander von Humboldt in the early nineteenth century. Humboldt laid no claim to having coined the term and used it without giving an exact defini-tion of what characteristics made a landscape a "steppe." While generations of scholars in different countries would refer back to Humboldt's use of the term, it was never entirely clear what exactly distinguished a steppe from other en-vironments and why the Russian grain-producing regions were usually denoted as steppes, but the environmentally similar Great Plains in the United States were not. Those who attempted definitions of the term around the turn of the century generally equated the steppe with other landscapes, attributing to it vastly divergent characteristics. The steppe could be, in order from more to less fertile soils, a form of grassland, heathland, or even the equivalent of a desert.[34]

When the geographer Robert Gradmann entered a turn-of-the-century debate on the history and human settlement of southern German landscapes, the term "steppe" was thus as ambiguous as ever. In his research into *Urland-schaften*, or "primordial landscapes," Gradmann postulated that forest and steppe periods had alternated in Europe with the cold conditions of recurring ice ages. He attempted a "negative" definition of steppes in opposition to forests. Still unconvinced, the desert geologist Johannes Walther intervened shortly after the First World War to call for a more limited, and thus more precise and useful, definition of the steppe that would distinguish it clearly from the desert. Just one year before that, the Russian-German climatologist Vladimir Köppen had attempted exactly that by introducing the "steppe" as an independent climatic zone in his famous classification scheme, which distinguished the steppe from the desert by its lower level of aridity.[35]

Over the 1920s, however, any clarity vanished once again. When the term was co-opted by *Ostforscher*, it became an ever more elusive and wide-ranging term, ultimately denoting any kind of barren and "degenerated" landscape. Despite his apparent attempts to objectively define the "steppe" before the First World War, Gradmann actually contributed to the further blurring of the term by inserting a racial dimension into his research, in which he connected certain styles of landscape and land use with particular racial characteristics. With this approach, he came close to Penck's research on *Kulturboden*, and in fact attended and contributed to meetings of the SdVK in the interwar period.[36]

In the 1930s, Gradmann incorporated the term *Kultursteppe*, or "cultural steppe," into his vocabulary, denoting the anthropogenic origin of steppes through destructive land use. As Gradmann himself stated in his writings, he borrowed the term from Friedrich Ratzel, who had already referred to the *Kultursteppe* of the East around the turn of the century. By including an explicitly evaluative dimension of environmental degradation, he loosened the "steppe" even further from a universal definition. And yet Gradmann was continuing his attempts to rein in the conceptual vagueness by following Walther's call and proposing another negative definition of the steppe through a contrast of its characteristics with the attributes of deserts. But Gradmann was no longer— or rather had never actually been—in control of the term. Over the 1920s, a growing number of *Ostforscher* referred to the steppes of the East; and after

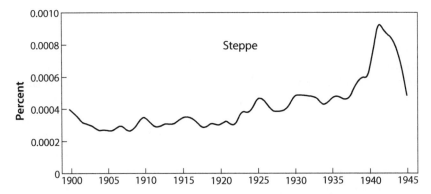

FIGURE 6.5. Frequency of the term "Steppe" in Germanophone literature in Google Books

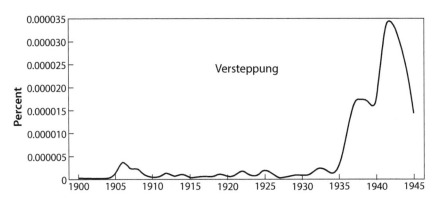

FIGURE 6.6. Frequency of the term "Versteppung" in Germanophone literature in Google Books

1933, the "steppe" became fully incorporated into the official propaganda about the deteriorated landscapes of the once fertile, ordered, and German East.[37]

The man who was responsible for the inflationary use of both the term itself and its close relative *Versteppung* (Figures 6.5 and 6.6) was Alwin Seifert, who—in an ironic turn—was actually focused on the German heartland rather than the mythical "German" landscapes of the East. Seifert had made a name for himself through his work as a landscape architect and "landscape advocate" in the construction of the Autobahn under Fritz Todt. Supported and protected by his powerful friend and employer, Seifert was able to propagate his controversial calls for both a new "organic" and holistic approach to engineering and a turn toward "biological-dynamic farming," a kind of organic farming developed by the founder of anthroposophy, Rudolf Steiner, and still practiced by his followers today.

Although Seifert's colleagues received many of his theories with skepticism and sometimes even open hostility, Nazi planners and landscape architects could not ignore his influential essays, which were later collected in *Im Zeitalter des Lebendigen*.[38] The most controversial—but also the most influential—essay in the volume was Seifert's 1936 piece on the "*Versteppung* of Germany," which made a big impression not only on Herman Sörgel but on engineers and planners all over the country. The fear of desertifying landscapes in Central Europe was not new: already in 1845 Humboldt had declared "dearth" as a ubiquitous and insidious issue in Prussia. Seifert himself had been writing on the effect of "climate deterioration" on the soil and on plant geography since the late 1920s and on desertification as early as 1932. In his first forays into the topic, Seifert was building on studies from the 1920s on "draining" German landscapes through modern water engineering and land reclamation projects—studies that were inspired both by new hydropower and river works and by the Weimar governments' attempts to cultivate marshes. In 1922, the German geologist Otto Jaekel, another student of the prolific teacher and Sahara traveler Karl von Zittel, had already referred to a "vicious circle of desiccation" that was threatening Germany.[39]

Other studies echoed this fear of inexorable climatic deterioration in Europe. The climatologist Paul Kessler had argued that desert-like phenomena were occurring outside arid zones even before the First World War. In the 1920s, he diagnosed a change of the climate toward continental conditions—that is, a more arid climate with larger temperature differences between summer and winter and between day and night. Kessler saw a general trend of the climate toward more arid conditions, but he ended with a stark warning about the increase of atmospheric carbon dioxide through the burning of coal, which could have "disastrous" effects for Russia and other continental areas. Other commentators, like the climatologist Richard Scherhag, continued to hypothesize periodic oscillations of global climatic conditions, citing warm winters in Europe, melting glaciers on Iceland, and the warming Labrador Current in the North Atlantic as evidence.[40]

Seifert's thinking did not originate nor vanish in an echo chamber. Although Seifert rarely cited the literature on climate change (or, for that matter, any literature), he proved to be well acquainted with the debate. He supported the hypothesis of widespread desiccation, looking as far afield as the alleged southward advance of the Sahara, growing aridity in southern Africa, desiccation of lakes in Russia, and land erosion in the United States, which had just been publicized by Paul Sears. In fact, the US Dust Bowl produced a large amount of interest, anxiety, and literature, which Joachim Radkau has credited with "epochal significance" for ecological thinking in the United States and beyond. Sears's 1935 book *Deserts on the March* became one of the most influential books on desertification and ecological disaster in the 1930s.[41]

Just like his colleagues abroad, Seifert struggled to identify a clear reason for the growth of deserts. While he referred to the Sahara's expansion as a consequence of deforestation by Africans, he implied that climatic changes had natural and anthropogenic causes. When writing about developments closer to home, Seifert seemed to follow the same logic. The signs were hard to miss, he argued in his 1935 article on *Versteppung*, that the aridity in Central Europe now resembled that of the Middle Ages. There was hard evidence, he continued, "that the steppe is advancing inexorably from the east to the southeast." Seifert insisted emphatically that this change in climate had nothing to do with the thirty-five-year climate cycles that Brückner had discovered: the climate change he referred to was bigger, enduring, and far more threatening.[42]

It was, however, not yet too late. Seifert was convinced that a new ethos of engineering could counteract at least the anthropogenic share of climatic changes and impede or even halt desertification. With reference to the Dust Bowl, which he already characterized as a man-made catastrophe, Seifert argued that modern engineering measures, including those of the Third Reich, had aggravated *Versteppung* in Europe as well. In private, Seifert directly condemned Nazi efforts to increase productivity at all costs in the so-called *Erzeugungsschlacht*, or the battle of production, which—according to Seifert—harmed the German landscapes. In a letter to Todt, Seifert also criticized the civilian drainage of moors in Germany for having created deserts. And even in public, Seifert remained combative and intractable, although he never opposed the political system of the Third Reich. In 1941, he stirred up controversy once again by asserting that German influence did not prevent environmental decline in the East. He even went as far as arguing that *Versteppung* was not only *happening* in regions under German influence but was, in fact, *particularly* bad there. While Todt remained loyal to his protégé Seifert even after this highly controversial argument, he did feel obligated to introduce Seifert's article with a number of caveats and moderating comments.[43]

At that point, Todt already knew how much trouble Seifert could cause. In the mid- to late 1930s, some German engineers and Nazi planners had taken Seifert's critique of engineering measures as an assault on their work and tried to discredit Seifert as a backward-looking technophobe out of touch with reality. Incensed by Seifert's critique of the economic measures of the Nazi government, even the minister of agriculture Darré—a fellow adherent of "organic" farming—joined the critics and derided Seifert's work as "wrongheaded, fantasy-ridden scribblings." The biggest point of contention was Seifert's claim of *Versteppung* within the borders of Germany. Seifert's 1935 article caused a long debate, which was memorialized in its own special volume. Todt, who edited the collection, attempted to clear the air but could not stop the persistent criticism of Seifert's theses. Seifert's detractors generally charged him with overstating his case. The water engineer Uhden described *Versteppung* as a "catch word" that did not survive closer scrutiny and also called into question the validity of Seifert's empirical examples. All of this criticism led Seifert to claim after the war that he had been the victim of an organized witch hunt trying to discredit his character.[44]

Seifert had at least somewhat of a point. He was not fundamentally hostile to technology, as his accusers claimed. Although Seifert did warn of the pitfalls and unintended consequences of modern engineering, he never proposed abandoning all technological means or returning to a preindustrial state of nature. He had, after all, contributed willingly to building the German highway system. The landscapes Seifert described were always cultural landscapes, which humans had been actively shaping with their tools at hand. In comments similar to not only the general tenor of Nazi planning but also the rhetoric of the German conservation movement, Seifert demanded a technology that would respect the conditions of the landscape. He commented favorably on hydropower's potential to cover the growing energy demand in Germany. And Seifert entertained plans for a grand transformation of the annexed and occupied territories in the East—plans on such a grand scale that they would be feasible only with large-scale engineering. In a 1940 essay about the "Future of the East German Landscape" Seifert shared his fellow landscape architects' enthusiasm about the "reclaimed" land and the potential for vast transformations, asking rhetorically, "Who could doubt that this would be the greatest glory that our generation can accomplish?"[45]

Seifert talked about planting hedgerows and the "Germanization" of the eastern landscape, but his essay remained rather cursory. Seifert was undoubtedly itching to be involved in the planning for the East, but he—like his outspoken critic Darré and other agencies in the chaotic assembly of competing offices in the Nazi bureaucracy—lost out in the internal power struggle for control over the new territories in the East. Seifert had brought the term *Versteppung* into the limelight, but it was now up to Himmler's office and, more specifically, the RKF planners to propose policies for reorganizing the East and refashioning desertified landscapes. It was not without a certain historical irony that Wiepking, Seifert's archenemy, ended up becoming a central figure in this planning process.

7

Eastern Deserts

CLIMATE AND GENOCIDE IN THE
GENERALPLAN OST

AMONG THE MOST infamous schemes of the Third Reich, the *Generalplan Ost*, or "General Plan East," stands out for its totalizing rationale that sought nothing short of a radical and all-encompassing transformation of landscapes and their inhabitants. In its own terms, the plan's creators aimed at "Germanizing" vast areas of what today are Poland, the Baltic States, Russia, and Ukraine—or the "bloodlands"—on the ill-defined border between Europe and Asia. The plan consisted of numerous memoranda on resettlement, genocide, economic development, and environmental transformation and was unprecedented in its intended disruption, transformation, and violence. The goal of refashioning environments and climates of potential settlement areas had already been present in François Roudaire's Sahara Sea and Herman Sörgel's Atlantropa—projects that, like the *Generalplan*, were also predicated on the belief that these large-scale transformations were technologically and organizationally feasible. And like Roudaire and Sörgel, Wiepking and his fellow RKF planners claimed the threats of desertification and climate deterioration as justification for their respective projects. The Nazi planners, however, went much further, exploiting the common anxieties over environmental decline to rationalize erasing the histories, the landscapes, and the people of occupied colonial territories.[1]

After the debate about Seifert's article had died down in the late 1930s, *Versteppung* persisted. Cited as the material evidence of simultaneous racial and moral degeneration, it now became a fully hybridized term that firmly connected environmental to cultural processes. This conflation of environmental and cultural decline had already been an element of the nineteenth-century debate on climatic changes in North Africa and had become even more pronounced in the early twentieth century—from Ellsworth Huntington writing on corresponding climatic and civilizational cycles to Herman Sörgel borrowing Oswald Spengler's ideas of an all-encompassing environmental and civilizational decline for his Atlantropa propaganda.

In the Second World War, this conflation took on a newly militarized form. The "Germanization of the soil" that Nazi planners attempted in the East was a question of population, agriculture, horticulture, and silviculture. To create new "healthy" German life in the annexed and occupied areas, these officials argued, the desertified land had to be turned into gardens, forests, and fields. RKF planners emphasized that only Germans—lauded as racially and culturally superior—could realize this enormous task. Only Germans would be able to continue and complete the work that Frederick the Great and his father had begun in the eighteenth century. For the *Ostland* planners, the encroaching steppe was thus a wake-up call and a legitimization for action, or as the National Socialist author Wilhelm Zoch put it in 1940, "the East does not so much cry out for people as it calls out for the German of Germanic blood."[2]

Wiepking's Planned Landscapes

The fortunes of the instigator of the *Versteppung* debate Alwin Seifert waned after the war began, and they ultimately crumbled when his mentor Fritz Todt died in a plane crash in 1942. At the same time, Heinrich Wiepking's professional career was taking the opposite course. While his first publications had revealed few signs of strong political convictions, Wiepking willingly worked within and for the Nazi bureaucracy after Hitler's rise to power. Having secured a professorship at the University of Berlin in 1934, Wiepking contributed to the new government in various functions—among them as part of the designer team for the Olympic Village in Berlin. He fully supported Nazi Germany's expansionist plans, despite some unpublished doubts about enlarging military grounds in the late 1930s, which, he argued, took away valuable land from German farmers. In 1939, after Germany invaded Poland, Wiepking immediately described German colonization in the East as "a task of top priority for our students." If he had intended to offer up his services to Nazi planning offices with the article, he had done so successfully. In 1941, Wiepking became an integral part of the planning bureaucracy, assuming a position as Konrad Meyer's "special deputy" in the RKF. Over the next few years, he would take on additional roles in the administration, such as the chairmanship of the group on "landscape maintenance in the new settlement areas" in the Forestry Ministry. From these positions of power, Wiepking could propagate his ideas about the need for an all-encompassing environmental redesign of the East.[3]

Despite their different professional fortunes during the war, Wiepking's and Seifert's views of the need and significance of landscape planning were surprisingly similar. "In the last century," Wiepking wrote in the SS magazine *Das*

Schwarze Korps, "more parts of the earth's surface have been desertified through overexploitation and human action than in all the preceding ages through genuine, natural processes." He believed that this could be reversed only through even more, but better organized and harmonized, human interventions into nature—a stance that was very close to Seifert's views on the use of technology. Despite their intersecting approaches, or maybe because of them, Wiepking and Seifert ended up becoming fierce enemies.[4]

The disagreement between them had started in 1931 with a trivial quarrel over a contract to design a private garden in Munich. After 1933, their animosity continued unabated, now touching on issues of personal standing and influence in the Nazi bureaucracy. The dispute took a personal turn with decidedly anti-Semitic undertones during a heated letter exchange in 1939, in which Wiepking and Seifert accused each other of Jewish connections and sympathies. The reason for the altercation was as complex as it was petty, including arguments over the value and work of other German landscape architects, disagreements over the cultural roots of garden design, allegations of cronyism, and recriminations over a case of alleged collusion in an election for the professional organization of horticulture in Germany. At one point, Seifert apparently even challenged Wiepking to a duel to be fought after the end of the war. The duel, however, never took place, and Seifert and Wiepking also found no other way to resolve their quarrel conclusively.[5]

Despite the ongoing row, Wiepking took Seifert's side in the debate on *Versteppung* in 1938, railing against the ignorance of "urbanized scientists" out of touch with the conditions of German landscapes. While occasionally referring to marshlands as signs of environmental degradation, Wiepking echoed Seifert's warning that water was often drained too rapidly from the land. When Seifert reminded his readers that "*Versteppung* was not just a question of the food supply but also a question of the soul," Wiepking warned in his writings that a desertified landscape would wear down "the creative human powers." Wiepking, like Seifert, had a complex relationship to technology and engineering. Despite advocating large-scale planning, he also reiterated the danger of certain kinds of unspecified technology, which, he claimed, had put the environment at peril. Like Seifert, Wiepking criticized urbanization, calling the modern city a "merciless desert." Also like Seifert, Wiepking used the Dust Bowl in the United States as an example to illustrate the dangers desertification posed to Europe. In *Versteppung,* Wiepking had found a congenial concept through which to legitimize his own designs of large-scale transformations of the environment. It would become an increasingly central leitmotif in Wiepking's work during the Third Reich. His 1942 article that warned readers of the "encroachment of the steppe" included maps indicating the westward move of desertification with thick black arrows protruding from the deep East somewhere beyond the borders of the illustration (Figure 7.1).[6]

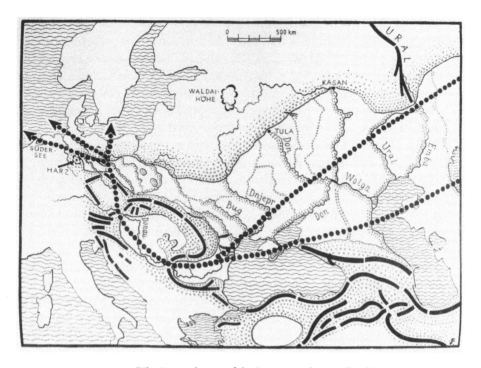

FIGURE 7.1. "The Encroachment of the Steppe into the Woodland," 1942.
Source: Heinrich Wiepking-Jürgensmann, "Das Landschaftsgesetz des weiten Ostens,"
Neues Bauerntum 34, no. 1 (1942): 9.

Through Wiepking's influence, *Steppe* and *Versteppung* also became central themes in the landscape-planning department of the RKF, where the terms appeared in memoranda and articles. Within Himmler's office, *Versteppung* and the associated concepts of climate change and soil erosion were becoming comprehensively politicized. In National Socialist planning circles, *Versteppung* became more than ever a sign of moral, cultural, and civilizational decline. This was particularly evident in the term *Kultursteppe*, Ratzel's description of anthropogenic desertification that Seifert propagated further in the Third Reich.[7]

True to his wholesale adoption of Seifert's conceptual framework, Wiepking elaborated on the moral dimension of the environment as well. In a turn of phrase reminiscent of Sörgel's belief in a link between environmental and cultural decline, Wiepking argued that not only the environment but "even the human mind can become desertified [*versteppen*]." To avoid this fate, the environment required the *Landschaftspflege* and *Klimapflege*, or "care of the landscape and climate," that Wiepking and his RKF colleagues advocated. Once again, Wiepking clearly connected the ability to provide this care to German *Kultur*, this time in the positive sense of the word and in contrast to the "calculating, mechanical, monistic, machine-like" civilization that had supposedly taken hold of Europe and was unable to generate "creative ideas."[8]

Yet Wiepking never lost his penchant for large-scale planning and engineering. He vocally opposed German nature conservationists, whom he considered unable to think holistically and thus unfit for the organizational tasks of planning and design. He even went so far as to claim that the organizational and ideological influence of nature conservationists on the Forestry Ministry was one of the foremost threats to the planning process in the East. Wiepking considered landscape planning to include large-scale engineering, which nature conservationists viewed with skepticism or even outright hostility. This positive view of human interventions into nature was also apparent in Wiepking's description of landscapes as "changeable" and "easily configurable." And he was not alone in holding these views; many among the circle of Nazi planners also believed in technological solutions on a large scale.[9]

Wiepking had always regarded himself more as a man of action than a theorizing academic. On assuming the professorship at the University of Berlin, he wrote to a friend that he saw his main role as strengthening the connection between "practice and the university." *Versteppung*, a topic with an academic background and an inherent call to action, gave him the perfect opportunity to do just that. Much like Seifert, Wiepking had started thinking about desertification in the context of his immediate surroundings. He described witnessing a movie scene set in the North African desert being filmed in the sand dunes of the Rehberge Park in Berlin. That it was even possible to model the Sahara in Germany, Wiepking wrote, could be regarded as evidence for an active process of desiccation. Before the Second World War, Wiepking also referred to the dry easterly winds as the "breath of the desert," causing climate deterioration. Toward the end of the 1930s, Wiepking's focus pivoted more and more to areas in the East. From a natural danger to all of Europe, desertification became a problem of the Polish settlement areas. And in Wiepking's reinterpretation of *Versteppung*—which he now seemed to equate with *Verostung*, or "Eastification"—this was no longer exclusively a natural problem but one both affected by and affecting human populations.[10]

During his work in the RKF, Wiepking brought *Versteppung*—and explanations of environmental processes in general—more and more in line with the racist core of Nazi ideology. In a 1939 article he described the historical achievements of Germans in the East whose hard-won fields he now saw as threatened by a "different world of foreign blood lurking in the pine bush, on sand and gravel, in marsh and moor." Picking up on the special relationship between Germans and nature that had been claimed and propagated in Germany during the 1920s and 1930s, Wiepking focused on the therapeutic or medical connotations of landscapes. According to the views presented in his *Landschaftsfibel* ("Landscape Primer"), landscapes directly expressed the qualities of their inhabitants: "healthy" landscapes reflected the virtues of strong races, whereas "diseased" ones were the effigy of "looting and nomadic" peoples. And the causality also worked the other way around: healthy landscapes could increase the strength of a people. In Wiepking's writings desertification, desiccation, and climatic changes, far from being neutral geological processes, became irrefutable evidence of racial or moral inferiority. They even offered a direct indication of the morality and behavior of their inhabitants: "The more neglected [*verwahrlost*] and degenerated a landscape is," Wiepking wrote, "the higher is the frequency of crime."[11]

In his writings, Wiepking both reflected and further inspired German ideas about the nature-culture—and climate-culture—connection, such as those of the botanist Erwin Aichinger who compared plant and human societies. Very much in tune with Wiepking's notions, Aichinger argued that just as vegetation could not prosper in the tundra, human development was also stifled in barren landscapes. But Wiepking also built on earlier theories of a link between climate, landscape, and national or racial characteristics that had been debated since the Enlightenment. In Germany, the psychologist and politician Willy Hellpach had taken up the topic in a new guise after the turn of the century, researching the psychological impact of climate and landscape characteristics. Hellpach focused increasingly on race as a category of analysis, developing the idea of *Volk*- or race-specific environments. In 1939, he called for "planned climate creation" to increase human well-being and productivity. Almost at the same time, RKF planners would define similar goals for their visions of climate engineering in the East, while emphasizing that the newly climate-controlled landscapes should be reserved for Germans only. This meant, too, that the East first had to be cleared of all non-Germanic populations.[12]

The *Generalplan* and the Climates of the East

During the Second World War, the RKF planners reflected Himmler's general obsession with projects on an ever increasing scale. And with the initially swift advance of German troops into the East, they had now seized the colonial land they needed to enact these plans. While the occupation of Poland in 1939 already called the planners to action, Hitler's decision to attack the Soviet Union in 1941 opened up unprecedented possibilities for colonial designs. In the elaboration of these designs, landscape planning became firmly connected to the far-reaching projects of ethnic cleansing and mass murder that the Nazis developed in the drafts for a comprehensive *Generalplan Ost*, which was to be the crowning achievement of their planning efforts in the East. The plan aimed to both destroy and develop on an unprecedented level: planners meticulously plotted the destruction of non-Germanic populations and their physical traces, preparing the ground for the creation of entirely new landscapes and climates.[13]

The name of the *Generalplan* first surfaced in 1939 or 1940. Konrad Meyer submitted the initial proposals to Himmler in 1940, and it was Meyer, as well, who quickly rose to become the most important manager of the plan. After a year of interministerial work, he sent the first version of the *Generalplan* to Himmler in July 1941. After this first version, the Himmler-led Reichssicherheitshauptamt (RSHA) worked on a second version, before Konrad Meyer received orders from Himmler in early 1943 to rework the *Generalplan* into a *Generalsiedlungsplan*, or a "General Settlement Plan." In all its iterations, the plan included far-reaching goals. As Meyer wrote in his editorial capacity for the volume *Landvolk im Werden*—a kind of companion guide to the *Generalplan*— "the aim of the settlement strategy is to Germanize the space down to the last little detail." This comprehensive "Germanization" required planning and engineering on an unparalleled scale. Like Wiepking, Meyer was critical of past uses of modern technology, but not of technology itself. With this approach, he reflected not just Wiepking's position but also that of the SS in general, which coupled technological optimism with premodern agrarian romanticism.[14]

Backed by the possibilities of modern engineering and technology, the *Generalplan* aimed at a "total solution" concerning the new organization of the East, mirroring Meyer's call for a "total perspective" of planning. Most prominently, these plans included the installation of German rule and a far-reaching colonizing agenda, envisioning the settlement of around 5.5 million German colonists in the occupied areas over twenty-five years. This also entailed displacing non-Germanic populations to Siberia or sending them to labor and death camps, as the land was to be "Germanized" comprehensively. A 1941 book on German rule in the East stated bluntly that no Pole or Jew was to remain on German soil. Complementary to these ethnic cleansing policies, the RKF plans for the East also included far-reaching designs for architectural, agricultural, and environmental transformation, which would ultimately serve as a model for reorganizing *all* landscapes in the new German Empire. In keeping with the general violence of Nazi rule, the plans to redesign the landscapes in the East called for prisoners and non-Germanic forced laborers to do the lion's share of the required work. Taken together, the *Generalplan* was thus part of a synchronized masterplan of ethnic cleansing, colonization, slave labor, and landscape transformation—exemplified in the luridly radical SS volume *Der Untermensch* ("The Subhuman"), which justified German rule and violence by alleging the "everlasting will of the subhuman to return the land to desert" and the German ability to "put his [positive] mark on the landscape."[15]

Nazi planners also conceived of the reorganization of East as a test for Germans to live up to the ideal of the "great colonizing *Volk*." The outcome of the plans would decide whether Germans deserved to occupy and retain the land in the East. They used this rhetoric to inflate the urgency and importance of the task. Once in charge as the "special deputy," Wiepking quickly became one of the main landscape planners in the RKF, involved in all aspects of the planning process. His designs for the East included new villages and towns that would comply with the German standards of a "green" and well-organized *Kulturlandschaft*, or "cultural landscape." Once again, this did not refer to a romantic notion of a stylized medieval village: "The good farmer," Wiepking wrote in 1943, "has always been a master of technology."[16]

Wiepking's main focus, however, remained on landscapes and the environment—the "source of *völkisch* power," where large-scale engineering and technology were just as important. His plans for the East included reclaiming fertile soil that had purportedly been covered over by moving sand. Wiepking targeted sandar (glacial outwash plains) and dunes that, in his assessment, threatened to silt up rivers and lakes and should therefore be removed. In a more holistic approach to *Versteppung*, he also advocated engineering a change of climate through creating artificial lakes and planting forests and hedgerows—the latter one of his pet projects throughout his lifetime (Figure 7.2).

"Rain is by no means made by God alone," he wrote in 1942, "even the climate is nothing else, after all, than the sum of the causes and effects of weather." Wiepking did have some doubts whether macroclimatic conditions could be completely controlled by humans. He believed nevertheless that they could at least be nudged in the right direction in an all-encompassing planning process: "In the East, we don't only need German people," he wrote, "but also trees, forests, clouds, and rain." Wiepking took part in an ongoing debate among Nazi officials about climate modification and amelioration that remained a key issue for planning offices throughout the war. He explicitly referred to the climate as changing and thus at least in theory adjustable. Most of the research on the possibilities of engineered climate change focused on eastern landscapes, such as the Ukrainian steppes. There the "special position" of technology in the East—particularly the centralization and authoritarian control over engineering projects—would allow for larger-scale approaches.[17]

Some German publications about climate control disregarded the usual anti-Slavic tenor and admitted that the Russians had already begun to

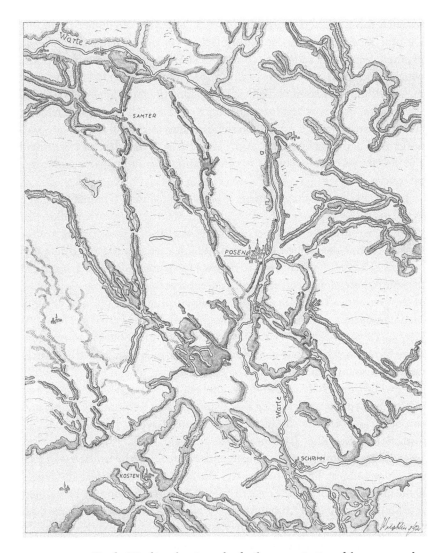

FIGURE 7.2. Map by Wiepking showing a plan for the reorganization of the area around Posen/Poznań, highlighting the use of hedgerows (Staatsarchiv Osnabrück).
Source: Niedersächsisches Landesarchiv Osnabrück,
K 2001/019 Nr. 96 H, n.d.

successfully combat aridity and drought, citing their "wind protection plantings," which RKF planners would often praise as staunchly German inventions. Doubts about the practicability of engineering climates nevertheless persisted. One report that was circulated among German offices in 1941 and 1942 stated that engineered climatic changes were possible but that more research was needed to gauge the practical potential of climate alteration. Despite these notes of caution and a lack of conclusive research, however, planners frequently called for climate modification of eastern landscapes, which the circular assessed as having a "very limited potential for development" without increased rainfall. Arguing along the same lines as Wiepking, the same report advocated for evaporation surfaces to increase precipitation and discussed the potential of subterranean dams to control artesian groundwater. And like Wiepking, the report also attributed the decrease of rainfall and the process of *Versteppung* to a combination of anthropogenic and "natural" causes.[18]

In this discussion about climate modification among German planners, Wiepking's RKF colleague and friend Erhard Mäding was another influential voice. He wrote about the impact of changes in the landscape on the macroclimate, or the *Gesamtklima*, of a region. Mäding even referred to the potential benefits of a control mechanism to regulate the amount of carbon dioxide in the atmosphere, alerting his readers to the effects of an anthropogenic increase of the greenhouse gas concentration from burning fossil fuels. In general, Mäding was very well-read in climatological matters. He cited the Russian-German climate scientist Wladimir Köppen and wrote about Wendler's studies on technological climate modification. He also mentioned Walther, Brückner, Penck, and Gradmann in his book on "Care for the Landscape." Directly borrowing from Hellpach, Mäding referred to climate engineering as an important step toward changing the "place-specific conditions of bodily and emotional [*seelisch*] life." In the East this meant, above all, the German transformation of desertified land into forests and fields.[19]

Attempts at modifying the climate, and transforming the landscape in general, did not stop at the well-being of the German settler. They were also linked to the military conflict on the eastern front. While reorganizing the East would actually divert resources from the military, planners endeavored to make their work appear essential to the war effort—especially in the last three years of the war. Wiepking, in particular, couched his plans in military language, putting "military-political information" above all other kinds of knowledge. Already in 1935, Fritz Wächtler had explored the connections between "cultural landscapes" and what he termed "defensive landscapes" in the military rearmament of Germany.[20]

Wiepking developed this idea into his concept of a *Wehrlandschaft*, or a "defensive" or "militarized" landscape—a concept that Himmler embraced with enthusiasm. In other documents, Wiepking referred to a *Kampfwald*, or a "fighting forest," without explaining what exactly this meant. It was clear from the descriptions, however, that *Wehrlandschaften* were wooded and green, in contrast to the wide-open spaces of the steppe. The bucolic appearance of the planned landscapes of the East would disguise military units, arms factories, and other military infrastructures. Wiepking's efforts to make landscape planning integral to war planning had its limits. Other Nazi offices sometimes questioned how the *Wehrlandschaft* would look and work in practice. When Wiepking argued for indigenous trees to be planted in Berlin because of their "military-political significance," for example, Albert Speer's office underlined the term with the pointed query "of a tree?" Nevertheless, *Wehrlandschaft* survived the debate and remained a term in the planning documents of the RKF.[21]

The interconnected ideas on *Versteppung*, climate engineering, and *Wehrlandschaften* all featured in the "Decree of the Commissioner for the Strengthening of Germandom 20/VI/42 on the Design of the Landscape in the Incorporated Areas in the East," which was to be a forerunner to an eventual Landscape Law. The directive, an elaboration of ideas only touched on in the *Generalplan*, represented the work of Wiepking with contributions by his RKF colleagues Meyer and Mäding and by Hans Schwenkel, a representative from the Forestry Office. The main idea behind the Landscape Decree was that there were race-specific environments that had to be kept free of impacts from alien races. This prerequisite implicitly called for ethnic cleansing in the East—a clear refutation of Wiepking's postwar claim that the Landscape Decree was the "most peaceful product of the Nazi period." Additionally, introducing race-specific environments meant calling for the remodeling of the eastern landscape to German standards. *Versteppung*—as a manifestation of the danger for the landscape from non-Germanic peoples—was the point of origin for all the rules, regulations, and measures to follow in the decree. The first two sentences read as follows: "Through the cultural ineptness of foreign *Volkstum* [nation/race], the landscape in the incorporated territories in the East is extensively uncared-for, barren, and desertified through acts of predatory exploitation. In large parts, it has taken on steppe-like characteristics."[22] The steppe had, by now, become the dominant description of the "deteriorated" eastern landscapes, overshadowing marshes and moors. In fact, the Landscape Decree made it clear that "water shall, as long as it is at all possible, be kept in the country."[23]

The directive continued with two of Wiepking's central projects—the proposals to plant hedgerows and forests to hold back the steppe, and those to create lakes and reservoirs to moderate the climate and increase precipitation in the East. The landscape, the document read, had to be transformed fully and irreversibly to become "intrinsically healthy, permanently full of life and harmonic" and thus an appropriate home for the new German owners and inhabitants. Like many Nazi plans, the directive remained obscure on the details. It revealed neither what exactly "intrinsically healthy" meant nor how this could be achieved. Preliminary designs of the RKF point to rather schematic and simplistic models of ideal landscape archetypes for the lands in the East that equated certain kinds of architectural styles and land use as "Germanic" and therefore good, omitting a closer look at local particularities.[24]

The Landscape Decree was Wiepking's most visible achievement in the RKF, but it was destined to remain unfulfilled and thus a side note in the history of the "German East": the decree—like the *Generalplan Ost* in general—was intended to be implemented fully after the final victory of the German troops over the Soviet Union. By the end of 1942, however, the tide of the war had clearly turned against Germany. This did not mean that the RKF stopped work. Initially, planning actually continued almost entirely unabated, and even in 1943 the German Research Foundation (DFG) still gave out money for research into the effects of tree and hedge plantings. The attempts at realizing the far-reaching plans also went ahead: the SS Commando Crimea, for example, carried on with their preparations for German settlement and explored the possibilities of active "climate control" until its last-minute evacuation in April 1944. And since the planning work of the RKF was classified as "essential to the war effort," the office continued with its designs for Germanizing the East until the Red Army was only a few weeks away from Berlin in 1945.[25]

The *Generalplan* in Action

Lauded as the primordial landscape of Germany, the forest had gained a deeply ideological and moral value in Nazi discourse over the 1930s. The Nazis built on earlier Romantic notions of a firm connection between the German nation and forests, but they also developed their own particular ideas. The forest, which they conceived of as a hierarchically ordered space with the ability for eternal renewal, became a symbol for the Nazi state and its ambition for an eternal, or at least thousand-year-long, *Reich*. Similarly, Nazi ideologists adopted the image of the constant evolutionary struggle of different species over forest resources as an allegory of the struggle of human races over *Lebensraum* and resources. Besides these ideological arguments, planners argued that wood was necessary for an autarkic German economy in preparation for war. In the 1930s, this strategic demand had fueled the efforts at afforestation in Germany, but it also led to the steadily increasing scale of clear-cutting. Wood was indeed becoming a more and more important energy resource toward the end of the war. Influenced by the moral and racial arguments of Nazi planners, however, literature on forests had come to subordinate the economic value of trees to their ideological importance in the racially charged fight for survival of the German *Volk*.[26]

It was thus unsurprising that one of the most urgent tasks for RKF planners in their fight against the eastern steppes was planting trees. They designed the resulting afforestation programs not only to give the landscape a more German character but also to aid in the climatic transformation of the annexed and occupied areas. In the first years of the war, RKF planners and their colleagues in the Forestry Ministry had already diagnosed a general scarcity of forests in the East—a finding of the extensive climatic-geographical survey of areas in the annexed areas to be settled by Germans. In the area around Zichenau/Ciechanów in southeast Prussia/Masovian Voivodeship, for example, the RKF estimated that only 10 percent of the area was forested—a state of affairs that planners portrayed as a consequence of the Polish overexploitation of the land.[27]

Overall, the RKF planned to plant more than ten thousand square kilometers of trees in the annexed areas in the East, which would have increased the forested area by more than 70 percent. In the end, the progress was rather pitiable, as the RKF managed to plant only about ten square kilometers per year. This was due to a variety of supply issues, among which the shortage of seeds and saplings for tree nurseries seemed to be the most pervasive. The afforestation program in Saybusch/Żywiec, which had the ambitious goal of transforming the "desertified [verwüstete] areas" into green, healthy landscapes, ran into serious supply issues as early as 1940. Compounding the slow progress of the program were communication issues between planners and local officials and a general lack of information on land tenure. The RKF was in constant contact with the Forestry Ministry over the difficulties of the tree-planting programs and called for cooperation among different offices involved in afforestation. The shared responsibility, however, did not lead to the desired results. Rather than a master plan, the offices produced many different, and often contradictory, piecemeal designs for afforestation. Throughout the German occupation, the progress of surveys and afforestation projects in the East remained hampered by supply issues and organizational shortcomings, which resulted not least from the ongoing disputes between the RKF and the Forestry Ministry over their respective spheres of influence and over different approaches to planning in the East.[28]

When the RKF attempted to carry out other parts of the Generalplan they ran into similar organizational issues. The RKF began forcibly displacing populations in Poland almost immediately after the Germans invaded. These rash and usually extremely violent actions destabilized the areas and made any other work even more difficult. In the area around Zamość in Poland, for example, the brutal removal of over 110,000 non-Germanic inhabitants to forced labor and concentration camps and their replacement with German colonists led to the long-lasting Zamość Uprising from 1942 to 1944. Yet Nazi planners were unfazed and continued their preparations for colonization further east—in the occupied areas of the Soviet Union. Between 1939 and 1943, about 1.25 million ethnic Germans were "resettled" in the East—most of them so-called *Auslandsdeutsche*, or "foreign Germans," with varying degrees of cultural affinity to Nazi Germany. The sheer numbers, however, could not conceal the fact that the colonization efforts did not go according to plan. The majority of the German settlers never found a new permanent home. They were simply moved from one temporary camp to the next. Moreover, German settlers were increasingly attacked by partisans. When their safety could no longer be guaranteed in the last two years of the war, Nazi authorities started to remove the settlers again. Even the RKF admitted as early as 1943 that the settlement attempts had been rushed. Local German officials in the East reflected that judgment, complaining about the negative consequences. One official even warned that the overpopulation—caused by Germans settling in the area—would lead to even more *Versteppung*. Here, the contradictions between RKF plans and the consequences of its practices became clearly visible.[29]

There was, in fact, general chaos: overlapping spheres of authority, contradictory directives from different offices, and a lack of coordination in the East. Mayer criticized the state of affairs as early as 1941, calling for a general overhaul of the bureaucracy. Wiepking—whose ambitious plans suffered under the chaos—criticized the situation as well, complaining to Meyer about the slow bureaucratic process of interministerial coordination. While Wiepking's *bête noire* was the Forestry Ministry, his colleague Mäding voiced his unhappiness about the rival offices of the Reichsstelle für Raumordnung (Office of Spatial Planning) and the Reichsstraßenverwaltung (Roadways Administration), which claimed authority over the planning process in the East as well. The bureaucratic chaos in planning for the East became so bad at one point that the drafts of the *Generalplan Ost* were lost for an entire year while being passed around from office to office. Faced with having to explain the lack of progress on the transformation of the East, planners now even used the "obstacle" of the purportedly steppified soils as an excuse for the slow amelioration process.[30]

Toward the end of the war and with the German army in retreat, planners encountered a whole new host of problems. Despite being classified as "essential to the war effort," landscape concerns were forced to take a backseat to the exigencies of the war. While the plans were as grandiose as ever, action to enact them slowly ground to a halt. This was not too surprising: sometimes, landscape planners' visions just ran completely counter to the needs of war, as in the case of Meyer's idea to use some vital agricultural land for afforestation. As the fortunes turned more and more against the Germans toward the end of war, the plans of the RKF for a new "Germanized" East were sometimes directly undone. All over the East, German officers and officials ordered the large-scale felling of trees for fuel and construction material, sometimes even in specifically protected areas.[31]

And yet the German planners still wreaked havoc in the occupied areas. While the RKF could not produce their vision of a German East, the chaotic population removals, afforestation programs, and reverberations of war caused large-scale destruction, violence, and death. In Belorussia alone, according to the historian Jürgen Zimmerer, a quarter of the population died, 30 percent lost their homes, industrial capacity fell by 90 percent, and the amount of fertile land was halved. After six years of the RKF fight against the steppes in the East, the region had not only been thrown into utter chaos at the cost of unimaginable human suffering, but had also become *less* fertile or—to put it into terms of the time—more *versteppt.*[32]

The Fate of *Versteppung*

During the Third Reich, the specter of an imminent *Versteppung* produced a program for environmental and climatic transformation of the East on a scale comparable to the anticipated changes of Atlantropa. Far from innocuous aesthetic interventions into the landscape, planting trees and creating lakes were integral parts of the grand design of creating a "German East," with all that the term entailed. Understanding clearly what stood behind the "resettlement" policy of Himmler's office, Wiepking and his fellow landscape architects and planners contributed willingly to the mass displacement and murder of Slavic and Jewish populations. The plans for landscape transformation were not just coincidentally connected to ethnic cleansing: planning on the scale that Wiepking envisioned was possible only in an "empty" colonial landscape, a kind of eastern frontier free of the supposedly destructive elements that were impeding German culture to take hold.

The alleged *Versteppung* of the East provided the physical manifestation of the destruction and an excuse for both population transfers and landscape planning. Wiepking was one of the most prominent promoters of the concept, borrowing not only from *völkisch* literature but also from the vocabulary that the climate change debate of the late nineteenth century had produced. Wiepking's colleague Mäding even drew direct parallels between the advance of deserts in the South and the advance of steppes in the East. This connection was common among planners, who referred to some parts of the East as *Wüstensteppengebiete* or "desert-steppe areas." At the same time, scientists continued to discuss large-scale climatic changes, including the threat of desertification and soil erosion in the East, which Nazi planners used in their arguments.[33]

Desertification had never been a politically neutral concept. European colonialists used it to legitimize their goals in North Africa from the 1870s, and

Sörgel used it to argue for his vision of Atlantropa in the 1920s. Among Nazi landscape planners, however, *Versteppung* took on an even more politicized and virulently racist tenor. Having moved far beyond a description of environmental change, *Versteppung* came to describe a cultural process in the double sense that the German term *Kultur* implied: desertification was the clear sign of both the corrupt relationship to the soil and the moral degradation that foreign non-Germanic races exhibited. In the RKF, and especially in Wiepking's writings, desertification became the prime justification for Germans to occupy and reorganize the annexed and occupied territories in the East. *Versteppung,* and with it a large part of climate science, moved from colonial science to colonial policy. Hermann Leiter, one of the strongest proponents of non-anthropogenic climate change in North Africa in the scientific debate, embodies this transition. In 1942, he wrote an official treatise on the value of land in the Ukraine for the "Great German" economy, describing the land optimistically as "fertile soil of the steppe," waiting for reclamation by German farmers.[34]

To become fertile, according to the rationale of RKF members, the steppe had to be conquered—first by German military occupation and then by large-scale planning. Just as the idea of desertification had a pre-Nazi history and did not emerge sui generis, so did the grandiose plans for environmental transformation. The historian Joachim Radkau has diagnosed a general "push towards bigness" among German engineers in the early twentieth century, who understood rationalization in terms of concentrating and coordinating technology. This penchant for large-scale technological systems and planning also inflected environmental engineering and landscaping. Wiepking claimed in the Landscape Decree that "the recreation of the eastern landscape is without antecedent." He was wrong. The *Generalplan* had antecedents in projects of water, landscape, and climate engineering even beyond Germany. After all, Wiepking's proposals to create immense standing bodies of water to effect anthropogenic climate change had an uncanny resemblance to François Roudaire's plans of the 1870s to inundate low-lying parts of the Algerian Sahara. And the frightening scope of Nazi plans for an entirely transformed environment in the East do not appear exceptional next to Hermann Sörgel's project to lower the Mediterranean Sea and irrigate the Sahara. Both Sörgel's project and the *Generalplan Ost* were strikingly similar in their utter disregard for people and their colonialist assumption of an empty, vast space—a tabula rasa—waiting for the "ordering hand" of the planner or engineer. While Atlantropa remained entirely in the sphere of unrealized visions and plans, the *Generalplan* and its companion plans were backed by the highest echelons of the Nazi bureaucracy and began to be enacted in the 1940s. Technology and large-scale engineering had already become part of *völkisch* discourses in the 1920s and became integral to Nazi ideology in the Third Reich.[35]

In the end, the defeat of the German military stopped Nazi planners from implementing the *Generalplan*, and with it, their attempts to transform the Eastern climate and landscape. It is important to note, however, that this was not the end of discussions about desertification, climate change, and climate engineering in Germany. First, there was no *Stunde Null* ("Zero Hour") in the field of landscape planning in postwar Germany. Both institutions and ideas survived the collapse of the Third Reich. Meyer, put on trial in Nuremberg, left the dock a free man and continued his academic work on planning in the Federal Republic. Wiepking went on to have a successful academic career in West Germany as well. As a witness at the Nuremberg Trials, he exposed the survival of old patterns of thought by arguing that the fight to fertilize the soil was now more important than ever considering that "over the last one-hundred years more land area has become desertified [*wüst*] than in the entire preceding history of mankind." And Wiepking's old nemesis Seifert, who would also enjoy a successful postwar career, had very similar ideas: "The world is not just coming apart at the seams politically," he wrote in 1948, "but also sociologically, morally, economically and generally, thus also meteorologically. We are at the dawning of a new age."[36]

The concept of *Versteppung* survived as well and even had a small renaissance in the second half of the 1940s. It still described the landscapes of the East, which remained under the threat of "advancing steppes," but now featured in books that mourned the "German homeland without Germans" in the East (Figure 7.3).[37] The postwar years in Germany also saw the belated publication of Nazi research on desertification conducted in Russia and the Ukraine. Anton Olbrich continued Wiepking's work on "wind protection plantings" in the Ukraine and still referred to the "cultural steppes" and "deserts made by human hands." In a softening of the cultural argument, Olbrich now held up Soviet measures against soil erosion as a good example that other countries, including Germany, should adopt. In the Soviet Union itself, grand plans to alter climatic conditions seemed strangely reminiscent of both the Nazis' professed anxieties of climatic deterioration and some of the Nazi schemes to engineer climates and environments. In Germany, other research on *Versteppung* maintained the tenor of the environmental threat from the Nazi era, but dropped the racist undertones: the desert ethnographer Findeisen warned of the "westward progress of desert climates" and the "wedge-like advance of steppes towards Europe" in 1950.[38]

With the Iron Curtain closing ever more tightly, however, the number of publications on the eastern steppe and *Versteppung* dropped sharply in the Federal

FIGURE 7.3. "Dying forest" in a shifting sand dune in Pomerania.
Source: Lutz Mackensen, ed., *Deutsche Heimat ohne Deutsche: Ein ostdeutsches Heimatbuch* (Braunschweig: G. Westermann, 1951), 9.

Republic. And even the Democratic Republic did not see another upsurge in steppe research. Academics in the Federal Republic reestablished their connections with scientists in the West. And in the period of decolonization, the focus of desertification research turned—once again—to the Sahara.

8

Epilogue

GLOBAL DESERTIFICATION AND
GLOBAL WARMING

AFTER THE SECOND WORLD WAR, climatologists remained uncertain about the causal mechanisms behind large climatic shifts and continued to disagree about the direction—and, more fundamentally, the existence—of global climate changes. Through the enduring discussions, changing climates had nevertheless become firmly established as a topic of scientific debate and research. Attempts at actively changing macro-climatic conditions, on the other hand, had recently experienced a serious setback—at least for the time being: the military defeat of Nazi Germany also entailed the end of their megalomaniacal plans to ethnically cleanse and climatically transform the landscapes in the East. The *Generalplan Ost* still stands as a cautionary tale of the utterly misguided technological hubris of Nazi planners. This, then, seems like a good endpoint to the story the preceding chapters have told.

And yet, it is difficult to identify an organic conclusion, when the interconnected histories of climate change ideas and climate-engineering schemes have continued and even gained momentum since the 1950s. In the second half of the twentieth century and in the early twenty-first, climate has taken on an even more public and political role than in the years covered in this book—and this calls for at least a short outlook on the trajectory of developments and dynamics within climatology and climate engineering in the more recent past. While my main goal in this book has been to explore the lesser known history of "changing climates" before the age of environmentalism and IPCC Reports, in the following pages I connect the storylines across the customary divide that cuts the Anthropocene into two halves—the one before the widespread realization of anthropogenic global warming and the one after. Rather than providing a full account of a history already told in more detail elsewhere, in this epilogue I highlight some of the connections and some of the breaks in the development of desertification research and climate engineering over the twentieth century.[1]

During the increasingly politicized debates on climate change, desiccation, and *Versteppung* of the early twentieth century, climatology was undergoing a gradual readjustment and reorientation. The new directions and methodologies in the field would ultimately open the door to the atmospheric physics- and computer-based climate science that developed in the second half of the twentieth century and is taught in universities today. Practitioners of this emergent "dynamic" climate science, growing out of Austrian and Scandinavian approaches around the turn of the twentieth century, continued to investigate climate variability and processes of climate change, but did so with a new focus on phenomena in the high strata of the atmosphere and air mass circulation. The climate historian Matthias Heymann has aptly described this development as part of a fundamental, if slow, methodological shift in climate research from inductive to deductive reasoning. Rather than inferring from local phenomena to global processes—as geographers and geologists had done in North Africa in the nineteenth century, for instance—the "new" climate science deduced local weather and climate conditions from the effects of global atmospheric processes, relying increasingly on global atmospheric circulation models (GCMs) and computer simulations.[2]

After the Second World War, a turn away from the heavily politicized debates about soil degradation and a general skepticism toward approaches used in Nazi geopolitics among scientists of all stripes amplified this methodological shift. The new methods did not conquer the whole field of climatology all at once. Rather, they led to a split in climate research that became apparent by the end of the 1960s. While global atmospheric phenomena like the Madden-Julian Oscillation, ENSO (El Niño Southern Oscillation), and anthropogenic global warming became main subjects of the new kind of climate science, desiccation and desertification—characterized as localized phenomena linked to particular land-use practices—remained in the realm of the telluric sciences of geography, geology, and soil science. The two strands of the debate on climate developed in surprising isolation from each other for several decades. They started to reconverge only in the 1990s, when the purely local explanations for desertification came under attack and the connections between global warming and local climate effects became an object of sustained research efforts.[3]

It was then, as well, that large-scale climate modification schemes enjoyed a renaissance, this time in the guise of geoengineering projects to halt and possibly reverse anthropogenic global warming. Even before that time, however, large-scale environmental engineering projects had not vanished from the scene. The public reckoning with the full scale of Nazi projects in the East had, in fact, little effect on the trajectory of environmental engineering, as the plans to Germanize the East were not commonly identified as central to Nazi

crimes. The belief among engineers and wide parts of the public in technological solutions to environmental problems continued unabated. And just as
popular anxieties over climate change persisted from the first half of the twentieth century, so did plans for engineering climates. In fact, the rise of atmospheric climate science inspired a new generation of engineers to come up
with new proposals to change weather and ultimately climatic conditions
through cloud seeding and other atmospheric interventions. Meanwhile, the
older notions of climate engineering through earthbound interventions continued to inspire large engineering schemes as well. Projects in Soviet Russia
to fertilize large stretches of arid land through irrigation and reservoir-induced
climate change brought Roudaire's and Sörgel's ideas about climate engineering into the second half of the twentieth century. And even the nineteenth-
century visions of a Sahara Sea have continued to reappear in various guises
up until the present day.

Desertification 2.0

While climate scientists became increasingly concerned with dynamic phenomena in the high atmosphere over the course of the twentieth century, the
debate on desertification became a subject of soil science, which focused on
land degradation and erosion. Rather than taking the increasingly global perspective of atmospheric circulation models, desertification studies focused on
particular places or regions, although researchers often assessed the threat of
land degradation to be global. In contrast to atmospheric climate science,
Western research into desertification also remained clearly connected to its
colonial roots. This was evident not only in the researchers' places of interest—
particularly newly independent African states—but also in the focus on non-
European land-use practices as potential sources of soil degradation and the
attendant power struggles over environmental control of the land.

Desertification discourses reappeared at the forefront of public and scientific debates in the late 1960s. This spike in interest was closely connected to
the Sahelian drought on the southern fringes of the Sahara, which resulted in
massive crop failures, famine, death, and dislocation. While it became quickly
evident that a decrease in annual rainfall was one of the main proximate causes
for the drought, the ultimate cause *behind* this shift in weather patterns—or
maybe even in climatic conditions—was less clear. In the 1970s, climate scientists had already found convincing evidence that the Sahel region had experienced recurrent droughts of ten- to twenty-year periods over the past few
centuries or more. But evidence about the historic water levels of Lake Chad
also suggested that the drought starting in the late 1960s was probably as least
as severe as any event in the last millennium. A recent article has called the

steady decrease of rainfall in the Sahel beginning in the 1950s "perhaps the most striking precipitation change in the twentieth-century observational record." This anomaly begged for an explanation, and early research into the causes of the drought blamed ongoing desertification. In turn, researchers at the time attributed this advance of hyperarid conditions to people and, in particular, to the increase of cattle and human populations in the Sahel and an ensuing overgrazing and agricultural overexploitation of the soil. While some researchers continued the colonial pattern of criticizing Africans for failing to care for their environment, others denounced the disruptive practices of colonial regimes for putting in place the dynamics that would ultimately cause the environmental damage.[4]

Whoever was singled out in particular, the scientific consensus in the 1970s was that human action was the root cause of soil degradation and thus of the ongoing desertification in the Sahel. This explanation built upon Stebbing's writings from the 1930s that had inspired Sörgel with their stern warning of an "encroaching Sahara" through land-use practices in Africa and beyond. Stebbing was also the mastermind behind the Anglo-French Forestry Commission, which conducted research in West Africa from 1936 to 1937. One of its members was none other than Aubréville, who would define "desertification" in 1949—thus providing scientists and the public with a useful term to frame their environmental theories and anxieties. It was not just this historical background, however, that influenced the international response to the Sahel drought in the 1970s. There was also a considerable amount of new scientific backing for the dogma of locally produced land degradation in desert environments.[5]

In 1974, Joseph Otterman presented his model of desertification, which would go on to become highly influential in the field. Using data from the Israeli-Egyptian border, he hypothesized that overgrazing would lead to a loss in vegetative cover, significantly increasing albedo. Otterman reasoned that this change of the earth's reflection coefficient would cause the soil surface to cool and reduce convective—or cloud-forming—processes and thus lead to a reduction in rainfall in the affected area. Just one year later, the famed meteorologist Jule Charney proposed a similar mechanism, postulating biogeophysical feedback loops that could perpetuate and augment desertification processes. This theory painted a rather alarmist picture of the situation in the Sahel, as it implied that desertification might continue in perpetuity or at least for a long time before the positive feedback loops were interrupted by some other force or mechanism.[6]

International policymaking institutions and in particular the United Nations quickly adopted claims that desertification was produced by humans and could potentially endure for long periods of time. The UN had been involved in climate and arid zone research from its founding and zeroed in on desertification as a main environmental issue in the 1970s. The United Nations Sahelian Office (UNSO) was created in 1973 in response to the drought, and one year later the UN General Assembly recommended measures to arrest desertification. This resulted in the Conference on Desertification (UNCOD) in Nairobi three years later. The proceedings of the conference warned that 33 to 43 percent of the earth's surface was already desert or semidesert and that desertification threatened 19 percent of the remaining area—distributed over more than two-thirds of the world's then 150 countries. In its description of the origins of desertification, the conference report mentioned larger climatic changes as a possible factor but made it clear where most of the blame lay, arguing that the equilibrium of nature "is readily disturbed when man makes use of the land."[7] This assessment led the delegates to create a Plan of Action to Combat Desertification, stressing that, above all, time was of the essence. If there is one central theme to the plan it is that action must not await complete knowledge about complex situations. The delegates recognized the need for immediate action in applying existing knowledge, not only to stop the physical processes of desertification but to educate people in minimizing the harm done to the fragile ecosystems of dry lands by existing economic and social activities.[8] Evidently, the Plan of Action called for far-reaching programs of land-use changes in regions declared to be undergoing or under threat of desertification *before* there was any conclusive evidence about the causes.[9]

Rather than bringing attention to the shortage of evidence, the 1977 conference thus helped to spread the notion that desertification was the direct consequence of local land-use practices. And because of the dearth of evidence, the UN also helped to universalize desertification, suggesting that it worked in similar or even identical ways in all semiarid regions across the world. The same rhetoric continued to feature heavily in the United Nations Environment Program's (UNEP) subsequent assessments of desertification and in the discussions at the UN Conference on Environment and Development in 1992. At that time, the UN also diagnosed a lack of any marked progress in the fight against desertification, which the conference report found to be advancing at virtually the same rate as fifteen years before—despite the lofty aims of UNCOD to completely stop desertification by 2000![10]

The bleak analysis in 1992 of the accomplishments of the 1977 Plan of Action ultimately spawned the UN Convention to Combat Desertification (UNCCD) in 1994. The UNCCD now recognized both "climate variability" and "human activities" as principal causes for desertification. And at least nominally the UNCCD's own plan of action paid closer attention to the principles of decentralization and local participation in its proposed measures against desertification. But while the UNCCD newly emphasized regional variety and local specificities, the policies they recommended remained nevertheless essentially biased toward global environmental management discourses. The UN's institutionalization of the desertification debate and the associated spread of reports and articles about *worldwide* environmental degradation over the preceding decades helped sustain this approach. With more research into the causes of desertification over the course of the 1990s, however, the scientific debate would become more locally specific and more globally aware at the same time. This seemingly contradictory development was due to critical reviews of generalized statements on the impact of land use on soil degradation, on the one hand, and to the incorporation of research on anthropogenic global warming, on the other.[11]

Around the same time that the UNCCD was founded in 1994, the scientific consensus on the causes of desertification that had existed since the 1970s began to break down. While some research on desertification confirmed a statistically significant overall increase in hyperarid areas in Africa over the course of the twentieth century, other studies became more skeptical that desertification was linear, or progressive, across the entire continent. Evidence from satellite images in the early 1990s complicated the picture by revealing periodic fluctuations of the Saharan border rather than the unidirectional desertification that Charney had foreseen and that the UN assessments had claimed. This led researchers to look at periodic environmental cycles in desert environments and the locally specific adaptations of human societies. While the resulting studies did not necessarily contradict the more general findings about how land-use practices impacted desertification, other studies began to question exactly that.[12]

Faced with the failure of UN measures against desertification and a rising number of local studies questioning the increasingly precarious scientific consensus, some researchers started to investigate alternative causal models. They paid particular attention to the potential links between the Sahelian drought and changes of the sea surface temperature (SST) of the oceans, which had been researched since the 1980s. Although scientists had not reached a conclusive answer about the origin of SST changes—and thus simply had moved the problem of causality to another realm—their shift of emphasis signaled a move away from the overwhelming focus on land-use patterns and the impact of local human-soil interactions. In the early 1990s, calls for a scientific reassessment grew louder. A 1991 paper on desertification put it bluntly, stating that so far "very little research has been carried out."[13]

Concurrently, other members of the scientific community started to criticize the lack of a clear definition of desertification, which, a 1993 paper argued, had become merely a "buzzword" without any real analytical value. One year later, the geographers David Thomas and Nicholas Middleton went so far as to ask whether desertification in general was simply a myth that scientists and policymakers had perpetuated by continually repeating the results stemming from one-sided and outdated research. The widely accepted belief that vegetation—and especially forests—could halt and reverse desiccation came under attack once again as well. In a 1999 article on the "desiccationist discourse" in twentieth-century India, Vasant Saberwal traced the tenacious notion of forests as multipurpose weapons against environmental and climatic decline to questionable and nonrepresentative research results from the 1920s, conducted primarily by the US Forest Service. Even at the end of the twentieth century, there was still no conclusive evidence about the alleged restorative effects of forests and no long-term data to detect clear directional trends of desertification in many regions said to be affected. The deforestation-desertification link, however, remained prevalent in the popular press due to the ongoing political valence of desertification—and thus land-right—issues and the power of bureaucracies, such as forestry departments, to shape the terms of the discourse.[14]

Other researchers in the 1990s argued in a similar vein that the quick jump to conclusions about the causes of desertification revealed a larger failure of the Western scientific community to understand African land use. On the one hand, this European and American "misreading" of landscapes was connected to the prevalence of equilibrium models of the environment in the 1970s, which posited the existence and desirability of steady-state environments. More and more research pointed to the inapplicability of the model to arid regions, which reflected the slow move of ecology away from the environmental equilibrium paradigm. On the other hand, the "misreading" was also an effect of Western scientists failing to develop any deeper insights into site-specific social contexts of African land use and their exaggeration of the scale of human-induced environmental changes, such as deforestation. These critiques led some in the scholarly community to call more loudly for interdisciplinary approaches in desertification research, which would combine approaches from the natural and social sciences.[15]

While popular understandings of desertification often still mirror the overly simplistic narrative from the 1970s, the scientific search for the causes of desertification has become much more complex today. Despite the ongoing difficulties of establishing direct causal links, researchers have compiled growing evidence over the past twenty-five years suggesting that local processes of desertification might, in fact, be connected to global atmospheric phenomena that local populations cannot control. In the early 1990s, scientists started to look into how global atmospheric circulation and global climatic phenomena might affect desertification in different regions of the world. In their research, they mainly followed two different—but not mutually exclusive—paradigms.[16]

On the one hand, climate studies started to expand on Charney's early research by introducing new ways to analyze land-surface-atmosphere interactions. Researchers combined models of bio-geophysical feedback loops with the most up-to-date GCMs, thus postulating a connection between local processes of land change and global processes of atmospheric circulation. This line of research was the foundation for the UNCCD's approach to considering both local human actions *and* climatic variability as potential causes for desertification. It also led to studies on the potential impact of temperature changes in the low atmosphere above desertified areas on larger, and even global, climatic patterns. A recent study has even suggested that local vegetation-environment feedbacks may trigger a cascade of amplifying effects from the local to the global, potentially producing critical changes in the large-scale climate system. While this remains a hypothesis to be examined in detail, the tendency toward connecting local environmental conditions to global atmospheric phenomena has been at the center of some recent approaches to studying desertification.[17]

On the other hand, climate scientists have investigated ocean-atmosphere interactions with updated GCMs, expanding on the research of SST changes conducted in the 1980s. This research has moved the focus from local environmental conditions to long-range and large-scale climatic phenomena. Recently, this latter approach, combined with findings from research into bio-geophysical feedbacks, has appeared to be the most promising avenue of research into the mechanisms behind desertification. The root causes of SST anomalies are still not conclusively explained—sulfate aerosol concentration, ocean variability, and discrete cooling events in the northern hemisphere have all been proposed as candidates. By now, however, researchers have firmly established the link between precipitation levels in the Sahara and a particular kind of SST anomaly.[18]

Irrespective of the complexities of recent explanatory models, a sizeable part of the latest desertification research has shown a tendency to go beyond a "normalized" local context. Some of the most convincing current research includes both environmental and cultural specificities of particular places and, at the same time, global atmospheric phenomena like the effects of anthropogenic climate change. This inclusive line of inquiry into desertification has not solved the mystery of the origins of drought conditions, let alone produced effective countermeasures. It has, however, led to the recent rapprochement between geographic and atmospheric approaches in climate science—a reconvergence of the two lines of inquiry in the field after their estrangement in the second half of the twentieth century. This development has also meant that desertification—long considered a locally produced, if globally occurring, phenomenon—has recently taken on a new "atmospheric" dimension, joining the "telluric" focus of soil science. And finally, the broader approaches in desertification research have produced new uncertainties about causal mechanisms and connections, resulting in a situation that may appear similar to the state of climatology around the turn of the twentieth century. The most immediate difference, however, is that scientists then had comparably little long-term data and few convincing models to explain climatic processes on a global scale, while scientists today arguably face a surplus of heterogeneous data and a wealth of complex models representing phenomena from local water yields to global atmospheric circulation.[19]

The Very Long Life of the Sahara Sea

In the context of the emerging modern environmental movement in the late 1960s, the concern over drought and desertification in the Sahel was not exclusively an issue for newly independent African governments and international organizations. With a new sense of the interconnectedness of environments and the limited resources on "Spaceship Earth," desertification became a public issue once again. The old anxieties about advancing deserts and the decline of environments returned to newspaper and magazine articles, which also buttressed the popular Western assessment of the desert as an undesirable and unproductive environment: writers cast the Sahara as a villain, invariably "marching" or "creeping" south into fertile territory and threatening to destroy the world's forests. This perceived danger emanating from arid environments led to a renewed interest among policymakers in designs to halt and reverse the advance of the desert, with proposals ranging from local projects of afforestation and irrigation to international and comprehensive plans of action, and to new and rekindled ideas for large climate engineering schemes.[20]

The fear that environments were generally unstable and declining had been a powerful driver of climate engineering projects from the 1870s to the 1940s, and it continued to exert a tenacious grip on the popular imagination. On the one hand, fears of a coming ice age resurfaced, based on calculations of the next global glaciation period that followed Milutin Milanković's findings from the 1920s and 1930s on orbital variations and their influence on incoming solar radiation. Proponents of a cooling hypothesis also relied on the availability of a wide variety of temperature records from around the world and in particular on an observed leveling off or even a fall of global temperatures from the 1940s onward. The assumption of a cooling trend in the Earth's atmosphere was still common among scientists in the 1970s, although it was based on insufficient evidence covering a relatively short time span. The notion of an imminent new ice age did not remain exclusive to academia: Lowell Ponte's 1976 book *The Cooling* expressed this idea in popular and vivid fashion, discussing both American and Russian projects of climate engineering and ruminating on global climate control. Just a year before that, however, Howard Wilcox had argued for the opposite development, a global warming trend, in *Hothouse Earth*. And the number of scientific articles on global warming multiplied around the same time as well. Just as in the late nineteenth century, scientific practitioners in the 1970s were split over which direction global temperatures were trending.[21]

The ongoing debates and anxieties over changing climates, and the persistent belief in technological solutions to climatic problems, helped to create an atmosphere conducive to public debates on engineered weather and climate change in the European and North American press. Additionally, the Second World War had spawned military investigations into weather and climate engineering. Researchers in the pay of military institutions in the East and West continued their wartime cloud-seeding experiments after the war, investigating both potential agricultural and military benefits of "fixing the sky." These attempts to change the weather—and possibly the climate—through atmospheric interventions were closely connected to the emergent field of computer-based numerical weather and climate analysis: after all, cloud seeding could be successful only if the movement of clouds and the consequent weather patterns could be predicted with appropriate accuracy. Moreover, numerical weather prediction had been self-consciously connected to designs of weather and climate control from the beginning: the mathematician John von Neumann, who had worked on the first computer-based weather forecast, justified the Meteorology Project at the Institute for Advanced Study in Princeton by arguing that it would take "the first steps toward influencing the weather by rational, human intervention." These ideas remained a big part of research programs in applied meteorology. A 1975 study warning of the potential harm of weather modification reported that over sixty countries had already experimented with it.[22]

Schemes for changing the climate through alterations of geographical and geological conditions survived as well. As in the late nineteenth century, planners were particularly active in desert environments, which—despite all the optimistic rhetoric from the turn of the twentieth century—had still not submitted to the predicted environmental and climatic transformations. North Africa remained one center of attention due to its closeness to Europe and France's continued colonial presence in Algeria. While their overseas empire disintegrated in the postwar years, some French commentators inflated the Sahara to the "land of the future," which could become the center of a smaller and possibly informal, but more powerful, technologically sophisticated, and better organized empire. This rhetoric received another push when oil was discovered in Algeria in 1957. A book from the same year described a wealth of new projects, ranging from irrigation measures to mining and oil drilling ventures that would completely transform southern Algeria into farmland with heavily industrialized centers. Institutional developments reflected the colonizing drive to develop the Sahara. On the occasion of the founding of the French-led Organisation commune des régions sahariennes (OCRS) in 1957, a presentation at the French Academy of Sciences celebrated the new society, with one author describing the development of the Sahara as the "greatest enterprise that France is presented with in the twentieth century."[23]

One of the projects that continued to inspire imitations was Roudaire's project of inland lakes in the Sahara. It was just when Sörgel's Atlantropa project was finally slipping into oblivion in the early 1950s that the French engineer Louis Kervran breathed new life into the dream of a Sahara Sea by founding the Technical Research Association for the Study of the Saharan Inland Sea. Kervran, today better known for his work on the unconventional theory of biological transmutation, thought of using the incline between the Mediterranean Sea and the bottom of the chotts for hydropower production. An article in a French popular science magazine described this new rendition of Roudaire's project as finally based on a "solid foundation." While Kervran's plans did not receive any serious consideration from the scientific or political community, they did engender a new wave of similar ideas and projects. And many of these projects stayed faithful to Roudaire's choice of the region of the chotts on the border between Algeria and Tunisia—sometimes enlarging the original plans by adding canals, tunnels, or other engineering extensions to the Sahara Sea, or even deliberating the filling of the chotts through a diversion of part of the Niger River toward the Gulf of Gabès. In a French article from 1955, the Atlantropa critic Kurt Hiehle spoke out against the successor projects to Roudaire's plans. He repeated a common criticism from the 1870s that the use of salt water would rather quickly return the flooded salt flats to their former state—just with even more layers of salt added.[24]

Despite these challenges, there were also strong defenders of the Sahara Sea's viability, although with some rather large modifications. With an allusion to American and Soviet experiments, a 1958 article suggested that applying thermonuclear energy via targeted subterranean explosions of nuclear bombs would make Roudaire's plans—and even an expansion of his plans—finally possible. From Roudaire's original plan of engineered climate change, however, the project had now become almost exclusively focused on creating an inlet into the Sahara to transport oil by tankers out of the desert. Just a year before the article appeared, the Association des recherches techniques pour l'étude de la mer intérieure saharienne—with the expedient acronym ARTEMIS—was founded to research the viability of an inland sea in the region of the chotts in order to develop access to the interior and establish a favorable microclimate in the region. The result of this assessment (if there ever was one) remains a mystery today. In any case, ARTEMIS itself was quickly becoming obsolete. The ongoing colonial crisis in Algeria and the eventual end of French colonial rule in 1962 threw a wrench in the plans to follow up on Roudaire's project and to develop the Sahara. The French leadership's new prestige projects now focused on continental France, such as building nuclear reactors to satisfy the country's rising energy demands without foreign coal or oil. This marked the end, or at least the temporary end, of the second wave of discussions about the Sahara Sea.[25]

In the 1970s, the Sahel Drought and new research into desertification shifted the debate on large desert engineering schemes from the northern to the southern regions of the Sahara. Against the background of anxieties over growing deserts—whether man-made or naturally occurring and whether locally or globally produced—large-scale environmental planning and engineering rose to prominence once again in the Western press. Even anti-desertification measures that did not have an explicit engineering component were conceived as all-encompassing radical measures that would lead to a complete environmental and social transformation across different regions and countries. In a 1974 feature in the *New York Times Magazine*, Martin Walker diverged from the paradigm of local causes for desertification in describing the Sahel drought as a consequence of a shift in weather patterns. While Walker consequently saw little hope for a quick fix to revert the Sahara's expansion, he called for a neocolonial "massive re-education project" that would "turn the nomads into settled farmers."[26]

Other commentators thought that the large-scale application of modern technology and engineering measures could solve the problem of desertification. A 1970s "Plan to Make the Sahara Bloom" used the possibly most overused cliché in desert engineering circles to endorse the use of a network of pumps to irrigate the Sahara. The scheme called for a pilot project, which would provide the blueprint for similar pumping stations on an ever larger scale, eventually forming an irrigation belt of two thousand miles from Senegal to Chad, which the backers believed would "halt the Sahara's onslaught." Using forest or vegetation belts to stop the advance of desert conditions had already been discussed in the nineteenth century and had then become the basis of Stebbing's proposals to halt the encroachment of the desert in the 1930s. The idea proved to have great staying power and continues to inspire programs in arid regions around the world. In China, the project to build a protective and country-spanning "Great Green Wall" against desertification from the north has been in progress since 1978. In Africa, as well, a Green Wall extending across the continent is still a topic of discussion today, although its exact form and location are yet to be determined. Whether forests will be the long-awaited remedy to stop the old menace of the desert or simply a case of "delusional development," as one former USAID worker has called them, also remains to be seen. In any case, the utility of forest belts will probably lie more in their ability to sequester carbon than in regulating hydrological conditions or physically stopping an expanding desert.[27]

The general concept of protective, anti-desertification belts did not exclusively focus on trees and vegetation. In the 1980s, the Belgian biologist turned UNESCO official Georges Hense had a far more technologically elaborate plan for building a very long salt water canal through the Sahara running from Mauritania via Niger and Chad to the Mediterranean coast at the border of Egypt and Libya. Hense proposed desalinization plants along the canal that would supply freshwater for irrigation. He also hoped that the canal would become a suitable environment for marine life that could serve as a new food source for Sahara dwellers. Besides unsolved technical difficulties, such as the canal's projected course passing over a mountain pass of five-hundred-meter altitude in the Tibesti mountain range in the central Sahara, the plan also did not account for the possibility that salt water could contaminate the freshwater reserves under the soil of the proposed canal. Unlike the plans of establishing forest belts, the idea of a trans-Saharan canal did not catch on, aside from inspiring a few isolated macro-projects of waterways in Africa.[28]

Around the same time of Hense's plans, however, Roudaire's project made yet another comeback, this time under the aegis of the independent countries of Algeria and Tunisia. In an attempt to strengthen bilateral relations, the two states signed a treaty in 1983 that included the creation of a society to study the possibility of creating an inland sea in the region of the chotts (SETAMI). One year later, a joint Tunisian-Algerian commission turned to the Swedish research group SWECO to conduct the necessary calculations that would touch on the meteorological, climatological, hydrological, hydrodynamic, agronomic, and economic aspects of the Sahara Sea. The results were devastating. SWECO estimated that the high costs of the project would be entirely out of proportion to the doubtful benefits; that the desired economic and agronomic gains of the project were uncertain; and—perhaps most importantly for this study—that the meteorological and climatic effects of the Sahara Sea would be negligible.[29]

SETAMI had thus received a crushing blow before it was even fully constituted, and the Algerian-Tunisian project never got off the ground. And yet the story of the Sahara Sea was still not quite over. In 2011, the reputable Springer publishing house printed an article on an "Artificial Gulf Formation Scheme" in Algeria and Tunisia. Among the three authors was the eccentric and highly prolific macro-engineering advocate Richard Cathcart, who had published a kind of biography of the "engeoneer" Hermann Sörgel in 1980 in which he called for a reevaluation of Sörgel's plans. Thirty years later, Cathcart took on Roudaire's plans in his and his coauthors' plans for CATS (the Chotts Algeria-Tunisia Scheme), which they described as a "landscape recovery macro-project" or—on a different page of the article and somewhat more confusingly—as a "doable dream macro-project." The proposed design includes the creation of Port Tritonis, a new large industrial port on the banks of the new Sahara Sea. And bringing the project firmly into the twenty-first century, the plans also anticipated using the chotts as an overflow basin to counteract the sea-rise level from global warming.[30]

Beyond the Sahara Sea

While Cathcart's reimagined visions of macro-engineering new bodies of water have, at least so far, not been realized, the number of desert lakes on the Arabian Peninsula has recently increased. Their presence, however, has not been the product of careful planning and did cause some bewilderment at first. As reported on NPR's *All Things Considered*, the lakes are actually an unintended consequence of modern water supply systems in the region: water from the ocean is desalinated, transported to inland cities, used in residents' bathrooms and kitchens, and—after treatment—used once again to water gardens and parks. The water then seeps into the ground and, in a manner similar to the creation of oases, percolates back up some distance away to form a lake. Close to the city of Al Ain in the United Arab Emirates, 150 miles from the sea, there is now a growing lake in between stretches of sand dunes. The lake has not (or not yet) had a marked effect on climatic conditions, but it has already influenced the local ecosystem. Some species, like the midas fly, have been displaced from the immediate area of the lake, while others—like birds and even fish—have found a new home in the middle of the desert. Whatever the eventual evaluation of this process may be, the local ecosystem is clearly undergoing a rapid change due to the new lake. As in most stories about human interventions that drastically alter environments, whether intentional or not, there are both benefits and costs that are not always easy to predict.[31]

At the scale of a mystery lake in the Arab desert, the changes may be rather limited and contained. While the midas flies may not return to the area to lay eggs, there are many vast stretches of desert that offer ideal conditions for the insects. And the increase in the number of birds in and around the lake—like herons, cormorants, and ferruginous ducks—could even be regarded as a positive step toward maintaining their populations, although the effect of the birds on the surrounding desert ecosystem remains to be seen. For people, the desalinized water from the sea makes it possible to settle, survive, and live in Al Ain, and the new unintended lake is merely an odd addition to the desert landscape and maybe even an aesthetically pleasing one at that. The long-term consequences of the lake are harder to predict with any degree of certainty. It is possible that the lake will raise local humidity levels and slightly reduce the aridity of the surrounding area. Judging from the consequences of irrigation measures in arid zones around the world, however, it is more likely that the leeching effect of the freshwater will mobilize salts in the soil, which will then gradually salinize the water and, ultimately, lead to the formation of a salt lake or a salt flat not unlike the chotts in southern Tunisia.[32]

Compared to these potential consequences of local interventions, the scale of the benefits and the attendant unintended consequences of global atmospheric engineering projects are incomparably larger. And these large-scale geoengineering projects feature heavily in today's discussions over how to mitigate global warming. While climatology has moved from the telluric to the atmospheric sciences over the twentieth century, so too have the visions of climate engineering become more and more concerned with fixing atmospheric rather than geological or geographic conditions. Geoengineering has, in fact, become almost synonymous with attempts to change parameters of the atmosphere, as in the National Academy of Science's definition of the term as "large-scale engineering of our environment in order to combat or counteract the effects of changes in atmospheric chemistry."[33]

Most of the geoengineering projects discussed today are aimed at counteracting the effect of increased levels of carbon dioxide and other greenhouse gases in the atmosphere. The umbrella term covers a number of approaches, which can be classified into two categories. The first focuses on removing carbon from the atmosphere through a variety of mechanisms, ranging from carbon sequestration through afforestation to fertilizing the oceans with iron. The second approach concentrates on managing solar radiation by reducing the impact of the sun in various ways—proposed projects range from injecting chemicals into the atmosphere to installing large mirrors in deserts or even in outer space. In both categories, proposals range from the "plausible to the absurd," as a New Yorker article has poignantly put it.[34]

Both efforts to reduce carbon dioxide in the atmosphere and efforts to reduce the impact of the sun's radiation can offer hope in a world whose leaders appear to be unable to agree on the ways, and even the necessity, to reduce greenhouse gas emissions. And even if there were a way to immediately reduce the emissions, the world would not be spared the potentially far-reaching effects of climate change. As the fifth IPCC Report predicts, most aspects of climate change will persist for many centuries even if carbon emissions were halted immediately. Against the backdrop of this rather gloomy outlook, the Harvard geoengineering advocate David Keith is making the case for geoengineering research and development and has invested money in carbon capture technologies. Hugh Hunt, a professor of engineering at the University of

Cambridge, has made a rather resigned case for the importance of the Strato-spheric Particle Injection for Climate Engineering Project (SPICE): "It isn't a cure for anything. But it could very well turn out to be the least bad option we are going to have."[35]

Proponents of geoengineering certainly have a point. An acceptance of the "least bad option," however, could also detract and divert resources from ongoing efforts to reduce greenhouse gas emissions and to develop new emissions-free or emissions-reduced technologies. And the risks of geoengineering, even if tested in models or on a small scale, are difficult to predict with any degree of certainty. This is true even with the newest generation of GCMs. While computer-based climate models have made great leaps in development over the past thirty years, they are not foolproof oracles. In fact, climate models now pose the predicament of either aiding a reductionist discourse on climate or otherwise becoming too complex to be practically useful. As contradictory as they seem at first, the two possibilities are in fact two sides of the same coin: it is true, as Mike Hulme has argued, that the predictive natural sciences have painted future scenarios that are almost exclusively determined by climate and climate-dependent forces. While these scenarios include the environmental changes wrought by an increase of global temperatures, they often do not account for the contingent aspects of social, political, economic, or cultural development that are the historian's bread and butter. At the same time, however, climate models that intend to incorporate these dimensions in a more holistic approach run into serious problems of data availability and commensurability and of problems of scale. The complexity of inclusive models also leads to a greater number of potential sources of error, which make it even more difficult to know what effects geoengineering interventions would have in practice.[36]

In his critical response to geoengineering, Jim Fleming, a historian trained in atmospheric sciences, has warned that to this day we still have to rely on back-of-the-envelope calculations and simple computer models. But even with complex climate models, geoengineering remains uncharted territory, opening the doors to a whole host of potential dangers, from militarization and a technocratic refashioning of international politics to profoundly changing human relationships to nature. In a statement in front of the US Congress in 2009, Fleming added: "Applied to geoengineering, we should base our decision-making not on what we think we can do 'now' and in the near future. Rather our knowledge is shaped by what we have *and have not* done in the past. Such are the grounds for making informed decisions."[37]

Fleming's statement is a fitting end point for this long epilogue or—in an alternative reading—for this rather short and cursory overview of the history of desertification research and climate engineering in the second half of the twentieth century. I would nevertheless like to conclude with a momentary return to the nineteenth century and to North Africa to draw out some tentative connections between the world of Barth, Fischer, and Roudaire and our present situation. In this study, I have referred to "climate change," "desiccation," and "desertification"—and sometimes I have even used the terms interchangeably, as their definitions were always in flux and overlapping. The large-scale climate change that scientists like Leiter thought about, however, was clearly different from what we call "global warming" today. Leiter was not sure of the causes of climate change, and in his analysis he focused on the earthbound environments of North Africa, rather than atmospheric gases and circulation models. And while anthropogenic warming did in fact take off among the rapidly advancing industrialization of the nineteenth century and was even beginning to be theorized back then, it revealed its full effect and became an accepted theory only in the second half of the twentieth century. Similarly, while Leiter and his colleagues already hypothesized global processes of desiccation and climatic changes, the idea of a global, interconnected environment—in which climate could become a phenomenon on a fully planetary scale—has fully blossomed only in the past fifty years.

And yet the anxieties and fears that potential climatic and environmental changes on a large scale evoked in the nineteenth century still sound very familiar to our ears. Fears about climatic catastrophes, even global ones, have a history that goes back much further than the second half of twentieth century, where they are usually placed in the context of the rising environmental movement and the discovery of global warming. This, however, is not to say that the anxieties then and now have the same basis or equal validity or even, as a malicious reading may have it, that today's fears are just another dramatic overreaction in a long history of environmental and climatic panics. The past is, indeed, a foreign country. While scientists around the turn of the twentieth century had widely different views on the existence and the potential mechanisms of large-scale climate change and could not come anywhere close to an agreement, today the mechanisms behind large-scale global climate change are backed up by a strong scientific consensus—stronger, in fact, than on most other current issues in science. Highly complex modern GCMs still run into problems of scale and are only as good as their often incomplete or heterogeneous datasets, but the drastic human effect on the atmosphere—and thus climatic conditions—is difficult to deny (although many still try to do exactly that). It is becoming increasingly clear as well that the vocal group of climate change deniers have been more concerned with politics and personal economic gain than with sound scientific practice.[38]

There are nevertheless some potential connections between then and now. For one thing, climate science is showing signs of becoming more methodologically inclusive again, increasingly considering not only atmospheric but also telluric factors of long-range climatic changes. This slow diversification of approaches in climate science, combined with the field's increased emphasis on the social and cultural dimensions of climatic change, provides fertile ground for interdisciplinary approaches, which may well include an important role for geographers. Beyond the internal disciplinary dynamics, there are also connections between the past and present of climate science that concern the societal context. Climate has not been transformed from a "complex scientific issue" to a "hot political issue" since the 1990s, as Amy Dahan has argued. Climate science and climate engineering were heavily political from the beginning and stayed that way from colonial to postcolonial and from national to international contexts.[39]

This, however, warrants not a general turn away from climate science and its findings as a whole, but rather close attention to research questions and practices in the field. Critically examining climate research is not the same as—and is indeed far more responsible than—refusing to engage with research findings outright. This is a point sometimes lost in the shrill public debate in the United States today about global warming. Other, more historical connections can also elucidate some of the dynamics around current discussions of climate change: some of the latest predictions of climatic trends indicate that the desiccation of North Africa that some scientists feared in the late nineteenth century is now actually happening in the twenty-first, driven by global warming. Against this background, it is not surprising that anxieties among policymakers and both local and international stakeholders over global desertification continue. While the climatological community has reached a consensus that global warming is both real and anthropogenic, the local and sociopolitical effects of climatic changes are still subjects of controversial debates. Even the most up-to-date climate models cannot accurately predict beyond a very close time horizon what effects the rise in global temperatures will have in particular places and regions. Research continues into the various potential mechanisms behind climate change, desiccation, and desertification in Africa and beyond. And with all of the remaining uncertainty over what exactly will happen and what kind of effect human interventions will have, the stories of climate change anxieties and of the search for safe climate engineering technologies will continue for the foreseeable future.

Introduction

1. Heinrich Barth, *Reisen und Entdeckungen in Nord- und Central-Afrika in den Jahren 1849 bis 1855*, 5 vols. (Gotha: Justus Perthes, 1857), 1:209–18; see also Steve Kemper, *A Labyrinth of Kingdoms: 10,000 Miles through Islamic Africa* (New York: Norton, 2012); on the Libyan rock art, see Francis L. van Noten, *Rock Art of the Jebel Uweinat* (Graz: Akademische Druck- und Verlagsanstalt, 1978); Jan Jelínek, *Sahara: Histoire de l'art rupestre libyen: découvertes et analyses* (Grenoble: J. Millon, 2004).

2. Barth, *Reisen und Entdeckungen*, 1:214–15.

3. Europeans played the most conspicuous role in the climate change and climate engineering debate in North Africa. There were similar debates about desiccation and climatic changes in the United States around the same time, but they did not take place in a similarly conducive institutional context. Meteorology and the related field of climatology did not become academic fields in US-American universities until the early twentieth century; see Kristine Harper, "Meteorology's Struggle for Professional Recognition in the USA (1900–1950)," *Annals of Science* 63, no. 2 (April 2006): 179–99.

4. Eduard Brückner, *Klimaschwankungen seit 1700, nebst Bemerkungen über die Klimaschwankungen der Diluvialzeit*, Geographische Abhandlungen 4 (Vienna: Ed. Hölzel, 1890), 34; the English translation is borrowed from Eduard Brückner, "Climate Change since 1700," in *Eduard Brückner: The Sources and Consequences of Climate Change and Climate Variability in Historical Times*, ed. Hans von Storch and Nico Stehr (Dordrecht: Kluwer, 2000), 115. Beattie gives a fitting definition of environmental anxiety as "concerns generated when environments did not conform to European preconceptions about their natural productivity or when colonisation set in motion a series of unintended environmental consequences that threatened everything from European health and military power, to agricultural development and social relations": James Beattie, *Empire and Environmental Anxiety: Health, Science, Art and Conservation in South Asia and Australasia, 1800–1920* (Houndmills, Basingstoke: Palgrave Macmillan, 2011), 1. On the significance of anxieties for colonial planning and colonial violence, see Mark Condos, *Insecurity State: Punjab and the Making of Colonial Power in British India* (Cambridge: Cambridge University Press, 2020); Amina Marzouk Chouchene, "Fear, Anxiety, Panic, and Settler Consciousness," *Settler Colonial Studies* 10, no. 4 (2020): 443–60; Harald Fischer-Tiné, *Anxieties, Fear and Panic in Colonial Settings: Empires on the Verge of a Nervous Breakdown* (Cham: Palgrave Macmillan, 2016).

5. On climate narratives, see the essays by Richard Hamblyn, Sverker Sörlin, Michael Bravo, and Diana Liverman introduced by Stephen Daniels and Georgina H. Endfield, "Narratives of

Climate Change: Introduction," *Journal of Historical Geography*, Feature: Narratives of Climate Change 35, no. 2 (April 2009): 215–22. On narratives of decline, see Peter Burke, "Tradition and Experience: The Idea of Decline from Bruni to Gibbon," *Daedalus* 105, no. 3 (1976): 137–52; Richard Grove and Vinita Damodaran, "Imperialism, Intellectual Networks, and Environmental Change: Origins and Evolution of Global Environmental History, 1676–2000," *Economic and Political Weekly* 41, no. 41–42 (October 2006): 4345–54, 4497–4505. On the cultural dimension of climate and climate change, see James Rodger Fleming and Vladimir Jankovic, "Introduction: Revisiting Klima," *Osiris* 26, no. 1 (January 1, 2011): 1–15.

6. Here, I am setting out to tackle Michael Osborne's call to investigate how "the colonial enterprise alter[ed] the deployment and content of science": Michael A. Osborne, "Science and the French Empire," *Isis* 96, no. 1 (March 1, 2005): 87. On the issues of "colonial science," see Helen Tilley, *Africa as a Living Laboratory: Empire, Development, and the Problem of Scientific Knowledge, 1870–1950* (Chicago: University of Chicago Press, 2011), 7–11. On the colonial dimension of climatology, see Deborah R. Coen, "Climate and Circulation in Imperial Austria," *Journal of Modern History* 82, no. 4 (December 1, 2010): 839–75; Deborah R. Coen, "Imperial Climatographies from Tyrol to Turkestan," *Osiris* 26 (2011): 45–65.

7. Cf. Fabien Locher and Jean-Baptiste Fressoz, "Modernity's Frail Climate: A Climate History of Environmental Reflexivity," *Critical Inquiry* 38, no. 3 (2012): 579–98.

8. On early ideas of climatic variability and desertification, see, e.g., Clarence J. Glacken, *Traces on the Rhodian Shore: Nature and Culture in Western Thought from Ancient Times to the End of the Eighteenth Century* (Berkeley: University of California Press, 1967), 659–63; Diana K. Davis, *The Arid Lands: History, Power, Knowledge*, History for a Sustainable Future (Cambridge, MA: MIT Press, 2015), 49–79; Lee Alan Dugatkin, "Buffon, Jefferson and the Theory of New World Degeneracy," *Evolution: Education and Outreach* 12, no. 1 (June 6, 2019): 15; Jean-Baptiste Fressoz and Fabien Locher, *Les révoltes du ciel: une histoire du changement climatique (XVe-XXe siécle)* (Paris: Éditions du Seuil, 2020); Richard H. Grove, *Green Imperialism: Colonial Expansion, Tropical Island Edens, and the Origins of Environmentalism, 1600–1860*, Studies in Environment and History (Cambridge: Cambridge University Press, 1995); Richard H. Grove, *Ecology, Climate and Empire: Colonialism and Global Environmental History, 1400–1940* (Cambridge: White Horse Press, 1997); Grove and Damodaran, "Imperialism, Intellectual Networks, and Environmental Change"; Kenneth Thompson, "Forests and Climate Change in America: Some Early Views," *Climatic Change* 3, no. 1 (1980): 47–64; Gregory T. Cushman, "Humboldtian Science, Creole Meteorology, and the Discovery of Human-Caused Climate Change in South America," *Osiris* 26, no. 1 (January 1, 2011): 16–44; Lydia Barnett, "The Theology of Climate Change: Sin as Agency in the Enlightenment's Anthropocene," *Environmental History* 20, no. 2 (April 1, 2015): 217–37; Lydia Barnett, *After the Flood: Imagining the Global Environment in Early Modern Europe* (Baltimore: Johns Hopkins University Press, 2019). On global desiccation research in the second half of the twentieth century, see Marc Elie, "Formulating the Global Environment: Soviet Soil Scientists and the International Desertification Discussion, 1968–91," *Slavonic and East European Review* 93, no. 1 (January 2015): 181–204.

9. Mary Louise Pratt, *Imperial Eyes: Travel Writing and Transculturation*, 2nd ed. (London: Routledge, 2008). More prosaically, the emergence of a "planetary consciousness" could also be described as the development of an "Earth system problem-framework," a term that Ken

Wilkening has used for his study on the history of the study of long-range transport of dust; see Ken Wilkening, "Intercontinental Transport of Dust: Science and Policy, Pre-1800s to 1967," *Environment and History* 17, no. 2 (May 2011): 313–39; Deborah R. Coen, *Climate in Motion: Science, Empire, and the Problem of Scale* (Chicago: University of Chicago Press, 2018).

10. Davis, *Arid Lands*; on the environmental history of North Africa and the Middle East, see, e.g., Alan Mikhail, ed., *Water on Sand: Environmental Histories of the Middle East and North Africa* (New York: Oxford University Press, 2013); Sam White, *The Climate of Rebellion in the Early Modern Ottoman Empire* (Cambridge: Cambridge University Press, 2013).

11. The political dimension of current global warming debates is discussed widely in the recent literature. For a particularly perceptive analysis, see Amy Dahan-Dalmedico and Hélène Guillemot, "Changement climatique: dynamiques scientifiques, expertise, enjeux géopolitiques / Climatic Change: Scientific Dynamics, Expert Evaluation, and Geopolitical Stakes," *Sociologie du travail* 48, no. 3 (July 1, 2006): 412–32; Amy Dahan-Dalmedico, "Climate Expertise: Between Scientific Credibility and Geopolitical Imperatives," *Interdisciplinary Science Reviews* 33, no. 1 (March 2008): 71–81.

12. Emil Deckert, *Die Kolonialreiche und Kolonisationsobjekte der Gegenwart: kolonialpolitische und kolonialgeographische Skizzen* (Leipzig: P. Frohberg, 1884), 116.

13. Georges-Louis Leclerc Buffon, *Histoire Naturelle, Générale, et Particulière*, 5 vols. (Paris, 1788), 5:244; cited in Locher and Fressoz, "Modernity's Frail Climate," 579.

14. The original German term used by Deckert is *korrigieren*.

15. On early modern projects of climate engineering, see Sara Olivia Miglietti, "Mastering the Climate: Theories of Environmental Influence in the Long Seventeenth Century" (PhD diss., University of Warwick, 2016); Anya Zilberstein, *A Temperate Empire: Making Climate Change in Early America* (New York: Oxford University Press, 2019); Sara Miglietti and John Morgan, eds., *Governing the Environment in the Early Modern World: Theory and Practice* (London: Routledge, 2017).

16. See, e.g., Bruno Latour, *We Have Never Been Modern* (New York: Harvester Wheatsheaf, 1993); Richard White, *The Organic Machine* (New York: Hill & Wang, 1995).

17. William Cronon, "Foreword," in *Mountain Gloom and Mountain Glory: The Development of the Aesthetics of the Infinite*, by Marjorie Hope Nicolson (Seattle: University of Washington Press, 1997), xii; David Edgerton, *The Shock of the Old: Technology and Global History since 1900* (Oxford: Oxford University Press, 2007), xv.

Chapter 1

1. For Marjorie Hope Nicolson's classic study on the reimagination of mountains as places of sublime beauty, see *Mountain Gloom and Mountain Glory: The Development of the Aesthetics of the Infinite* (Seattle: University of Washington Press, 1997). On the Romantic fascination with glaciers, see Robert Macfarlane, *Mountains of the Mind: A History of a Fascination* (London: Granta, 2003), 103–36. On the connections between mountaineering and glaciology, see Garry K. C. Clarke, "A Short History of Scientific Investigations on Glaciers," *Journal of Glaciology*, Special Issue (1987): 4–24; Bruce Hevly, "The Heroic Science of Glacier Motion," *Osiris* 11 (January 1, 1996): 66–86. On the Romantic fascination with desert environments, see Cian Duffy, *The Landscapes of the Sublime 1700–1830: Classic Ground* (Houndmills, Basingstoke:

Palgrave Macmillan, 2013), 135–73; Uwe Lindemann, *Die Wüste: Terra incognita, Erlebnis, Symbol: Eine Genealogie der abendländischen Wüstenvorstellungen in der Literatur von der Antike bis zur Gegenwart* (Heidelberg: C. Winter, 2000), 112–44. On the cultural history of wastelands, see Vittoria Di Palma, *Wasteland: A History* (New Haven, CT: Yale University Press, 2014); Diana K. Davis, *The Arid Lands: History, Power, Knowledge*, History for a Sustainable Future (Cambridge, MA: MIT Press, 2015).

2. Arthur Stentzel, "Die Ausdorrung der Kontinente," *Naturwissenschaftliche Wochenschrift* 4, no. 45 (1905): 712–16. For a small selection of the literature on desiccation and desertification research and anxieties, see Gregory T. Cushman, "Humboldtian Science, Creole Meteorology, and the Discovery of Human-Caused Climate Change in South America," *Osiris* 26, no. 1 (January 1, 2011): 16–44; Clarence J. Glacken, *Traces on the Rhodian Shore: Nature and Culture in Western Thought from Ancient Times to the End of the Eighteenth Century* (Berkeley: University of California Press, 1967); Richard H. Grove, *Green Imperialism: Colonial Expansion, Tropical Island Edens, and the Origins of Environmentalism, 1600–1860*, Studies in Environment and History (Cambridge: Cambridge University Press, 1995); Richard H. Grove, *Ecology, Climate and Empire: Colonialism and Global Environmental History, 1400–1940* (Cambridge: White Horse Press, 1997); Richard Grove and Vinita Damodaran, "Imperialism, Intellectual Networks, and Environmental Change: Origins and Evolution of Global Environmental History, 1676–2000," *Economic and Political Weekly* 41, no. 41–42 (October 2006): 4345–54, 4497–4505.

3. Louis Agassiz, *Études sur les glaciers* (Neuchâtel: Jent et Gassmann, 1840), 225.

4. James Rodger Fleming, *Historical Perspectives on Climate Change* (New York: Oxford University Press, 1998), 11–20; Franz Mauelshagen, "The Debate over Climate Change in Historical Time, c. 1750–1850," *History of Meteorology* (forthcoming); Fredrik Albritton Jonsson, "Climate Change and the Retreat of the Atlantic: The Cameralist Context of Pehr Kalm's Voyage to North America, 1748–51," *William and Mary Quarterly* 72, no. 1 (January 1, 2015): 99–126; Edward Gibbon, *The History of the Decline and Fall of the Roman Empire*, 2nd ed., vol. 6 (London: Strahan and Cadell, 1788), 519; Roberto M. Dainotto, *Europe (in Theory)* (Durham, NC: Duke University Press, 2007), 84–86; Martin J. S. Rudwick, *Bursting the Limits of Time: The Reconstruction of Geohistory in the Age of Revolution* (Chicago: University of Chicago Press, 2005); Ivano Dal Prete, "'Being the World Eternal . . .': The Age of the Earth in Renaissance Italy," *Isis* 105, no. 2 (2014): 292–317. The monumental change in conceptions of the age of the earth can be traced through visual representations of time: Anthony Grafton and Daniel Rosenberg, *Cartographies of Time* (New York: Princeton Architectural Press, 2010). On the development of ice age theories in the eighteenth and nineteenth centuries, see Tobias Krüger, *Discovering the Ice Ages: International Reception and Consequences for a Historical Understanding of Climate* (Leiden: Brill, 2013).

5. Louis Agassiz, *Des glaciers, des moraines et des blocs erratiques, discours prononcé à l'ouverture des séances de la Société helvétique des sciences naturelles, à Neuchâtel le 24 Juillet 1837* (Neuchâtel: Société helvétique des sciences naturelles, 1837); see also Jean-Paul Schaer, "Agassiz et les glaciers. Sa conduite de la recherche et ses mérites," *Ecologae geologicae Helvetiae* 93, no. 2 (2000): 233–34; Martin J. S. Rudwick, *Worlds Before Adam: The Reconstruction of Geohistory in the Age of Reform* (Chicago: University of Chicago Press, 2008), 517–34.

6. Jean-Baptiste Joseph Fourier, "Extrait d'une mémoire sur le refroidissement séculaire du globe terrestre," *Annales de chimie et de physique* 13 (1820): 418–33; Jean-Baptiste Joseph Fourier,

"Remarques générales sur les températures du globe terrestre et des espaces planétaires," *Annales de chimie et de physique* 27 (1824): 136–67. The most complete expression of Lyell's uniformitarian beliefs can be found in Charles Lyell, *Principles of Geology; Being an Attempt to Explain the Former Changes of the Earth's Surface by Reference to Causes Now in Operation*, 3 vols. (London: J. Murray, 1830). For an account of Agassiz's unwavering opposition to theories of evolutions, see Edward Lurie, "Louis Agassiz and the Idea of Evolution," *Victorian Studies* 3, no. 1 (September 1, 1959): 87–108; Louis Agassiz, "Observations sur les glaciers," *Bulletin de la Société géologique de France* 9 (1838): 443–50.

7. HLHU MS Am 1419, Series I: Agassiz to Buckland, n.d. [probably 1838], 107. Agassiz's theories prefigured the more recent "Snowball Earth" hypothesis, which was first coined in 1992; see Joseph Kirschvink, "Late Proterozoic Low-Latitude Global Glaciation: The Snowball Earth," in *The Proterozoic Biosphere: A Multidisciplinary Study*, ed. J. William Schopf and Cornelis Klein (Cambridge: Cambridge University Press, 1992), 51–53. For an overview of the "Snowball Earth" hypothesis, see Paul F. Hoffman and Daniel P. Schrag, "Snowball Earth," *Scientific American* 282 (2000): 68–75. For a summary of the effects of the glacial theory on the main lines of geological thought around the middle of the nineteenth century, see Dennis R. Dean, *James Hutton and the History of Geology* (Ithaca, NY: Cornell University Press, 1992), 248–51.

8. With a good dose of self-confidence, Kelvin announced that his results invalidated all uniformitarian principles, proposed by Lyell and championed by a large part of the geological profession at the time. See William Thomson Kelvin, "The 'Doctrine of Uniformity' in Geology Briefly Refuted," *Proceedings of the Royal Society of Edinburgh* 5 (1865): 512–13. For a more in-depth discussion of Kelvin's calculations of the age of the earth and the following debates, see Stephen G. Brush, *The Temperature of History: Phases of Science and Culture in the Nineteenth Century* (New York: B. Franklin, 1978), 29–44; Joe D. Burchfield, *Lord Kelvin and the Age of the Earth* (Chicago: University of Chicago Press, 1990); Frank D. Stacey, "Kelvin's Age of the Earth Paradox Revisited," *Journal of Geophysical Research: Solid Earth* 105, no. B6 (2000): 13155–58. On Fourier's influence on Kelvin, see T. Mark Harrison, "Comment on 'Kelvin and the Age of the Earth,'" *Journal of Geology* 95, no. 5 (September 1, 1987): 725–29; Crosbie Smith, "Natural Philosophy and Thermodynamics: William Thomson and 'The Dynamical Theory of Heat,'" *British Journal for the History of Science* 9, no. 3 (November 1, 1976): 305–9.

9. William Thomson Kelvin, "On the Age of the Sun's Heat," ed. David Masson, *Macmillan's Magazine* 5, no. 29 (1862): 393; see also William Thomson Kelvin, "On the Secular Cooling of the Earth," *Transactions of the Royal Society of Edinburgh* 23 (1864): 160. In the 1880s, Kelvin attempted to square his theory of cooling celestial bodies with the evidence of past ice ages; see Burchfield, *Lord Kelvin and the Age of the Earth*, 41–42. The idea that the sun might eventually stop emitting light and heat had already been a topic of discussion before the nineteenth century: Frank A. J. L. James, "Thermodynamics and Sources of Solar Heat, 1846–1862," *British Journal for the History of Science* 15, no. 2 (July 1, 1982): 155–56.

10. John Tyndall, "On the Absorption and Radiation of Heat by Gases and Vapours and on the Physical Connexion of Radiation, Absorption and Conduction," *Philosophical Transactions of the Royal Society of London* 151 (1861): 1–36; Fourier, "Remarques générales sur les températures du globe terrestre et des espaces planétaires"; Jean-Baptiste Joseph Fourier, "Mémoire sur les températures du globe terrestre et des espaces planétaires," *Mémoires de l'Académie royale des sciences* 7 (1827): 569–604. For a portrayal of Tyndall and his scientific work, see Ursula DeYoung,

A Vision of Modern Science: John Tyndall and the Role of the Scientist in Victorian Culture (New York: Palgrave Macmillan, 2010); Roland Jackson, *The Ascent of John Tyndall: Victorian Scientist, Mountaineer, and Public Intellectual* (Oxford: Oxford University Press, 2020); Svante Arrhenius, "Über den Einfluss des atmosphärischen Kohlensäuregehalts auf die Temperatur der Erdober-fläche," *Bihang till Kongl. Svenska Vetenskapsakademiens Handlingar* 22, no. 1 (1896): 1–102—the English version appeared as Svante Arrhenius, "On the Influence of Carbonic Acid in the Air upon the Temperature of the Ground," *Philosophical Magazine* 41, no. 251 (1896): 237–76. See also Elisabeth Crawford, "Arrhenius' 1896 Model of the Greenhouse Effect in Context," *Ambio* 26, no. 1 (February 1, 1997): 6–11; Elisabeth T. Crawford, *Arrhenius: From Ionic Theory to the Greenhouse Effect* (Canton, OH: Science History Publications, 1996). On alternative hypotheses about the cause of ice ages, among them James Croll's astronomical theory (which would later serve as the basis for Milutin Milanković's work on orbital variation), see Krüger, *Discovering the Ice Ages*, 399–440; Svante Arrhenius, "Naturens värmehushållning," *Nordisk tidskrift* 14 (1896): 11.

11. Agassiz, *Études sur les glaciers*, 237–38.

12. Adolphe D'Assier, "Glacial Epochs and Their Periodicity," *Scientific American Supplement*, no. 632 (February 11, 1888): 10097–98; the article originally appeared in French as Adolphe D'Assier, "Les époques glaciaires et leur périodicité," *Revue scientifique*, 3, 24, no. 18 (October 29, 1887): 554–60. Hermann Fritz, *Die wichtigsten periodischen Erscheinungen der Meteorologie und Kosmologie* (Leipzig: Brockhaus, 1889); Charles Austin Mendell Taber, *The Cause of Warm and Frigid Periods* (Boston: George H. Ellis, 1894). There is now a very robust literature on the Little Ice Age; for some examples highlighting the social, economic, and political consequences, see Wolfgang Behringer et al., *Kulturelle Konsequenzen der "Kleinen Eiszeit"—Cultural Consequences of the "Little Ice Age"* (Göttingen: Vandenhoeck & Ruprecht, 2005); Geoffrey Parker, *Global Crisis: War, Climate Change and Catastrophe in the Seventeenth Century* (New Haven, CT: Yale University Press, 2017).

13. Macfarlane, *Mountains of the Mind*, 128; Henry H. Howorth, *The Glacial Nightmare and the Flood: A Second Appeal to Common Sense from the Extravagance of Some Recent Geology* (London: Sampson Low, 1892); John Tyndall, "On the Conformation of the Alps," *Philosophical Magazine* 4, 28, no. 189 (1864): 264; Peter H. Hansen, *The Summits of Modern Man: Mountaineering after the Enlightenment* (Cambridge, MA: Harvard University Press, 2013), 180–86.

14. For more on Ruskin's role in shaping Victorian ideas about society, the environment, and climate, see Vicky Albritton and Fredrik Albritton Jonsson, *Green Victorians: The Simple Life in John Ruskin's Lake District* (Chicago: University of Chicago Press, 2016).

15. For a complete account of Barth's voyage, see Steve Kemper, *A Labyrinth of Kingdoms: 10,000 Miles through Islamic Africa* (New York: Norton, 2012). The term "heroic science" is borrowed from Hevly, "Heroic Science of Glacier Motion." On the Romantic origins of the figure of the explorer, see Carl Thompson, *The Suffering Traveller and the Romantic Imagination* (Oxford: Oxford University Press, 2007); Gustav von Schubert, ed., *Heinrich Barth, der Bahnbrecher der deutschen Afrikaforschung* (Berlin: D. Reimer, 1897), iii. On the nationalist dimension of nineteenth-century German exploration, see Matthew Unangst, "Men of Science and Action: The Celebrity of Explorers and German National Idenity, 1870–1895," *Central European History* 50, no. 3 (2017): 305–27.

16. Mungo Park, *Travels in the Interior Districts of Africa: Performed under the Direction and Patronage of the African Association, in the Years 1795, 1796, and 1797* (London: W. Bulmer, 1799);

Mungo Park, *The Journal of a Mission to the Interior of Africa, in the Year 1805*, 2nd ed. (London: J. Murray, 1815); Friedrich Hornemann, *Tagebuch seiner Reise von Cairo nach Murzuck, der Hauptstadt des Königreichs Fessan in Afrika, in den Jahren 1797 und 1798*, ed. Carl König (Weimar: Landes-Industrie-Comptoir, 1802). For more information on Hornemann's journey, see Jos Schnurer and Herward Sieberg, eds., *F. K. Hornemann (1772–1801): "Ich bin völlig Africaner und hier wie zu Hause": Begegnungen mit West- und Zentralafrika im Wandel der Zeit* (Hildesheim: Universitätsbibliothek Hildesheim, 1999); Heinrich Barth, *Travels and Discoveries in North and Central Africa: Being a Journal of an Expedition Undertaken Under the Auspices of H.B.M.'s Government, in the Years 1849–1855*, 5 vols. (London: Longman, Brown, Green, Longmans & Roberts, 1857); Heinrich Barth, *Reisen und Entdeckungen in Nord- und Central-Afrika in den Jahren 1849 bis 1855*, 5 vols. (Gotha: Justus Perthes, 1857). On the Royal Geographical Society, see Felix Driver, *Geography Militant: Cultures of Exploration and Empire* (Oxford: Blackwell, 2001), chap. 2.

17. Humboldt to Barth, February 26, 1859; cited in Rolf Italiaander, ed., *Heinrich Barth: Er schloß uns einen Weltteil auf. Unveröffentlichte Briefe und Zeichnungen des großen Afrika-Forschers* (Bad Kreuznach: Pandion, 1970), 155.

18. Heinrich Barth, *Wanderungen durch das punische und kyrenäische Küstenland oder Mâg'reb, Afrikia und Barka* (Berlin: Wilhelm Hertz, 1849), 1:425–26 and 504–7; Barth, *Reisen und Entdeckungen*, 1:216. Barth was correct in his assessment of environmental and climatic changes in North Africa. The more humid conditions, however, had lasted only through the early and middle Holocene (up to approximately 6,000 to 4,000 years BP); see Alfred Thomas Grove and Oliver Rackham, *The Nature of Mediterranean Europe: An Ecological History* (New Haven, CT: Yale University Press, 2001), 209–20.

19. Paul Ascherson, "Reise nach der Kleinen Oase in der Libyschen Wüste im Frühjahr 1876," *Mitteilungen der Geographischen Gesellschaft in Hamburg* 1–2 (1876–77): 68. On the life and work of Gerhard Rohlfs, see Anne Helfensteller and Helke Kammerer-Grothaus, eds., *Afrika-Reise: Leben und Werk des Afrikaforschers Gerhard Rohlfs, 1831–1896* (Bonn: PAS, 1998); Erwin von Bary, "Über den Vegetationscharakter von Aïr. Schreiben des Dr. Erwin v. Bary an Prof. P. Ascherson," *Zeitschrift der Gesellschaft für Erdkunde zu Berlin* 13 (1878): 351–55; Gerhard Rohlfs, "Zur Charakteristik der Sahara," *Zeitschrift der Gesellschaft für Erdkunde zu Berlin* 14 (1879): 368–74.

20. Gerhard Rohlfs, *Quer durch Afrika: Reise vom Mittelmeer nach dem Tschad-See und zum Golf von Guinea* (Leipzig: F.A. Brockhaus, 1874), 110; Karl Alfred von Zittel, *Die Sahara: Ihre physische und geologische Beschaffenheit* (Kassel: Theodor Fischer, 1883), 38–39. See also Theobald Fischer, "Zur Frage der Klima-Änderung im südlichen Mittelmeergebiet und in der nördlichen Sahara," *Petermanns Mitteilungen aus Justus Perthes' Geographischer Anstalt* 29 (1883): 3–4. Theobald Fischer, *Studien über das Klima der Mittelmeerländer*, Ergänzungsheft zu Petermanns Geographischen Mitteilungen 58 (Gotha: J. Perthes, 1879), 44; Hermann Leiter, *Die Frage der Klimaänderung während geschichtlicher Zeit in Nordafrika*, Abhandlungen der k. k. Geographischen Gesellschaft in Wien 8 (Vienna: R. Lechner, 1909), 95; Duveyrier and Nachtigal also continued Barth's search for desert paintings and engravings; see Paul G. Bahn, *The Cambridge Illustrated History of Prehistoric Art* (Cambridge: Cambridge University Press, 1998), 45.

21. The figure of the heroic explorer played an important part in European scientific culture of the nineteenth century, which placed a high value on firsthand experience and eyewitness accounts; see Driver, *Geography Militant*. The same was true for scientists who worked in and

on glacial environments; see Hevly, "Heroic Science of Glacier Motion." On the importance of go-betweens for information gathering in imperial contexts, see Simon Schaffer et al., eds., *The Brokered World: Go-Betweens and Global Intelligence, 1770–1820* (Sagamore Beach, MA: Science History Publications, 2009); Dane Kennedy, *The Last Blank Spaces: Exploring Africa and Australia* (Cambridge, MA: Harvard University Press, 2013), 159–94; Kapil Raj, "Go-Betweens, Travelers, and Cultural Translators," in *A Companion to the History of Science*, ed. Bernard Lightman (Chichester: Wiley-Blackwell, 2016), 39–57. On the issue of a balanced representation of interactions between the colonial landscape, Europeans, and the indigenous population, see Andrew Sluyter, *Colonialism and Landscape: Postcolonial Theory and Applications* (Lanham, MD: Rowman & Littlefield, 2002), 11–27, 211–32.

22. Gerhard Rohlfs, *Reise durch Marokko, Uebersteigung des grossen Atlas. Exploration der Oasen von Tafilet, Tuat und Tidikelt und Reise durch die grosse Wüste über Rhadames nach Tripoli*, 4th ed. (Norden: Hinricus Fischer Nachfolger, 1884), 118–19, 256; Rohlfs referred to the "pushing back" of the Arabs, but his image of an all-European Algeria and his conviction that "there will always be some peoples who have to make room for others for the improvement of humanity as a whole" clearly point to the ultimate implications of his ideas.

23. Leiter, *Die Frage der Klimaänderung während geschichtlicher Zeit in Nordafrika*, 99; Wilhelm R. Eckardt, *Das Klimaproblem der geologischen Vergangenheit und historischen Gegenwart* (Braunschweig: F. Vieweg und Sohn, 1909), 122. Despite the criticism, reports by missionaries remained an important source for the debates on climate change in Africa; see Georgina H. Endfield and David J. Nash, "Missionaries and Morals: Climatic Discourse in Nineteenth-Century Central Southern Africa," *Annals of the Association of American Geographers* 92, no. 4 (December 1, 2002): 727–42.

24. Cf. Matthias Heymann, "Klimakonstruktionen," *NTM Zeitschrift für Geschichte der Wissenschaften, Technik und Medizin* 17, no. 2 (May 1, 2009): 171–97.

25. On the colonial context for the development of early climatic theories and concepts, see Grove, *Green Imperialism*; Grove, *Ecology, Climate and Empire*. For a critique of Grove's theses, especially his claim that an ecological mindset, including ideas of conservation and sustainability, developed in the colonies, see Joachim Radkau, *Natur und Macht: Eine Weltgeschichte der Umwelt* (Munich: Beck, 2000), 198–201. On the colonial origins of desiccation research, see Georgina H. Endfield and David J. Nash, "Drought, Desiccation and Discourse: Missionary Correspondence and Nineteenth-Century Climate Change in Central Southern Africa," *Geographical Journal* 168, no. 1 (March 1, 2002): 33–47. On the ecological history of the Sahara, see M. A. J. Williams, *When the Sahara Was Green: How Our Greatest Desert Came to Be* (Princeton, NJ: Princeton University Press, 2021); William R. Thompson and Leila Zakhirova, eds., *Climate Change in the Middle East and North Africa: 15,000 Years of Crises, Setbacks, and Adaptation* (New York: Routledge, 2022).

26. On the development of geography as a colonial science, see David N. Livingstone, *The Geographical Tradition: Episodes in the History of a Contested Enterprise* (Oxford: Blackwell, 1993); Anne Godlewska and Neil Smith, eds., *Geography and Empire* (Oxford: Blackwell, 1994); Morag Bell, Robin A. Butlin, and Michael J. Heffernan, eds., *Geography and Imperialism, 1820–1940* (Manchester: Manchester University Press, 1995); Driver, *Geography Militant*. On the Petermann publishing house, see Heinz Peter Brogiato, "Gotha als Wissens-Raum," in *Die*

Verräumlichung des Welt-Bildes: Petermanns geographische Mitteilungen zwischen "explorativer Geographie" und der "Vermessenheit" europäischer Raumphantasien, ed. Sebastian Lentz and Ferjan Ormeling (Gotha: Klett-Perthes, 2000), 15–29. On ecological imperialism, see Peder Anker, *Imperial Ecology: Environmental Order in the British Empire, 1895–1945* (Cambridge, MA: Harvard University Press, 2002). On the development of climatology as an imperial science, see David N. Livingstone, "Climate's Moral Economy: Science, Race and Place in Post-Darwinian British and American Geography," in *Geography and Empire,* ed. Anne Godlewska and Neil Smith (Oxford: Blackwell, 1994), 132–54; Richard H. Grove, "Imperialism and the Discourse of Desiccation: The Institutionalization of Global Environmental Concerns and the Role of the Royal Geographic Society, 1860–1880," in *Geography and Imperialism, 1820–1940,* ed. Morag Bell, R. A. Butlin, and Michael J. Heffernan (Manchester: Manchester University Press, 1995), 36–52; Katharine Anderson, *Predicting the Weather: Victorians and the Science of Meteorology* (Chicago: University of Chicago Press, 2005), 231–84; Martin Mahony, "For an Empire of 'All Types of Climate': Meteorology as an Imperial Science," *Journal of Historical Geography* 51 (2016): 29–39; Ruth A. Morgan, "Climate and Empire in the Nineteenth Century," in *The Palgrave Handbook of Climate History,* ed. Sam White, Christian Pfister, and Franz Mauelshagen (London: Palgrave Macmillan, 2018), 589–603; Philipp Lehmann, "Average Rainfall and the Play of Colors: Colonial Experience and Global Climate Data," *Studies in History and Philosophy of Science Part A,* Experiencing the Global Environment 70 (August 1, 2018): 38–49. On acclimatization science in the nineteenth century, see Michael A. Osborne, "Acclimatizing the World: A History of the Paradigmatic Colonial Science," *Osiris,* 2nd Series, 15 (January 1, 2000): 135–51.

27. Fleming, *Historical Perspectives on Climate Change,* 33–44; Paul N. Edwards, "Meteorology as Infrastructural Globalism," *Osiris,* 2nd Series, 21 (January 1, 2006): 229–50. On the development of nineteenth-century meteorological observation in France and Germany, see Fabien Locher, *Le savant et la tempête: étudier l'atmosphère et prévoir le temps au XIXe siècle* (Rennes: Presses universitaires de Rennes, 2008); Klaus Wege, *Die Entwicklung der meteorologischen Dienste in Deutschland* (Offenbach am Main: Deutscher Wetterdienst, 2002). The Austrian climatologist Julius Hann authored the standard work of the field; see Julius von Hann, *Handbuch der Klimatologie,* 1st ed. (Stuttgart: J. Engelhorn, 1883); for the English translation, see Julius von Hann, *Handbook of Climatology,* trans. Robert DeCourcy Ward (London: Macmillan, 1903). On Hann's contributions to modern climatology, see Peter Kahlig, "Some Aspects of Julius von Hann's Contribution to Modern Climatology," in *Geophysical Monograph Series,* ed. G. A. McBean and M. Hantel, vol. 75 (Washington, DC: American Geophysical Union, 1993), 1–7.

28. Herodotus, *Herodot's von Halikarnaß Geschichte,* trans. Adolf Schöll (Stuttgart: J. B. Metzler, 1828), 1:548ff. Whether Herodotus had actually traveled to North Africa, rather than relying on secondhand information, is in doubt; see Detlev Fehling, "The Art of Herodotus and the Margins of the World," in *Travel Fact and Travel Fiction: Studies on Fiction, Literary Tradition, Scholarly Discovery, and Observation in Travel Writing,* ed. Zweder von Martels (Leiden: Brill, 1994), 1–15. On the accounts of Barth's reliance on Herodotus, see Henri Paul Eydoux, *L'exploration du Sahara* (Paris: Gallimard, 1938), 11–12. Herodotus's account of Lake Tritonis in North Africa would become very important for François Roudaire's project of a Sahara Sea, which is described in detail in the following chapter.

29. Fleming, *Historical Perspectives on Climate Change*, 16; Diana K. Davis, *Resurrecting the Granary of Rome: Environmental History and French Colonial Expansion in North Africa* (Athens: Ohio University Press, 2007); Andrea E. Duffy, *Nomad's Land: Pastoralism and French Environmental Policy in the Nineteenth-Century Mediterranean World* (Lincoln: University of Nebraska Press, 2019). On the lasting importance of the "Granary of Rome" narrative, see Yves Lacoste, *Ibn Khaldun: The Birth of History and the Past of the Third World* (London: Verso, 1984).

30. It is, indeed, still a topic of debate today, although the consensus view is that, despite a somewhat wetter climate in North Africa and the Near East during Roman times, climatic conditions were similar to those in the nineteenth century before the onset of anthropogenic global warming; see H. H. Lamb, *Climate, History and the Modern World*, 2nd ed. (London: Routledge, 1995), 157. See also Grove and Rackham, *Nature of Mediterranean Europe*, 142–43. Issar, in contrast, argues that there were global warming events around 4000 BP, 3800 BP, and 1400 BP, which particularly affected arid and semiarid regions in the Mediterranean basin: Arie S. Issar, "The Impact of Global Warming on the Water Resources of the Middle East: Past, Present, and Future," in *Climatic Changes and Water Resources in the Middle East and North Africa*, ed. F. Zereini et al. (Berlin: Springer, 2008), 145–64. See also A. Issar and Mattanyah Zohar, *Climate Change: Environment and Civilization in the Middle East* (Berlin: Springer, 2004).

31. Eckardt, *Das Klimaproblem der geologischen Vergangenheit und historischen Gegenwart*, vii.

32. Davis, *Resurrecting the Granary of Rome*; see also Andrea E. Duffy, *Nomad's Land Pastoralism and French Environmental Policy in the Nineteenth-Century Mediterranean World* (Lincoln: University of Nebraska Press, 2019).

33. Carl Nikolaus Fraas, *Klima und Pflanzenwelt in der Zeit. Ein Beitrag zur Geschichte beider* (Landshut: J. G. Wölfe, 1847), 41–49; see also Radkau, *Natur und Macht*, 160–64; Oscar Fraas, *Aus dem Orient* (Stuttgart: Ebner & Seubert, 1867), 1:213–16.

34. Davis writes that it was after the 1853 publication of Ernest Carette's study on the origins and principal tribes of North Africa that French writers placed the blame for the alleged environmental degradation of the region on the Arab invasion of North Africa in the seventh century: Davis, *Resurrecting the Granary of Rome*. See also Ernest Carette, *Recherches sur l'origine et les migrations des principales tribus de l'Afrique septentrionale et particulièrement de l'Algérie*, vol. 3, Exploration scientifique de l'Algérie (Paris: Imprimerie impériale, 1853); on the rise of desicationist thinking among foresters, geographers, and natural scientists working in the British and French colonial tropics after the 1860s, see Grove and Damodaran, "Imperialism, Intellectual Networks, and Environmental Change," 4346–49.

35. The German lack of benevolence toward African populations was more than proven by racist legislation in the colonies and the colonial genocide of the Hereros in Southwest Africa in 1904; see Jürgen Zimmerer and Joachim Zeller, eds., *Völkermord in Deutsch-Südwestafrika: Der Kolonialkrieg (1904–1908) in Namibia und seine Folgen* (Berlin: Links, 2003); Isabel Hull, *Absolute Destruction: Military Culture and the Practices of War in Imperial Germany* (Ithaca, NY: Cornell University Press, 2005). On German colonial climatology in Southwest Africa, see Harri Olavi Siiskonen, "The Concept of Climate Improvement: Colonialism and Environment in German Southwest Africa," *Environment and History* 21, no. 2 (2015): 281–302.

36. On Theobald Fischer's research, see Karl Oestreich, "Theobald Fischer. Eine Würdigung seines Wirkens als Forscher und Lehrer," *Geographische Zeitschrift* 18, no. 5 (January 1, 1912): 241–54. While Fischer's climatological work is largely forgotten, Braudel cited him among the most important contributors to climatological studies in the Mediterranean region; see Fernand Braudel, *The Mediterranean and the Mediterranean World in the Age of Philip II*, trans. Richard Lawrence Ollard, vol. 1 (Berkeley: University of California Press, 1995), 267–75; Fischer, *Studien über das Klima der Mittelmeerländer*, 41–42.

37. See, e.g., Eduard Brückner, *In wie weit ist das heutige Klima konstant? Vortrag gehalten auf dem 8. Deutschen Geographentage zu Berlin* (Berlin: W. Pormetter, 1889), 102; Henri Schirmer, *Le Sahara* (Paris: Hachette, 1893), 121; Wilhelm Sievers, *Afrika: eine allgemeine Landeskunde* (Leipzig and Vienna: Bibliographisches Institut, 1895), 167. While Fischer himself did not venture an explanation for this process, he continued to be fascinated by the issue of expanding deserts until the end of his life and helped to inspire geological research into desert formation; see Theobald Fischer, "Über Wüstenbildung," *Dr. A. Petermann's Mitteilungen aus Justus Perthes' geographischer Anstalt* 57 (1911): 132; Lino Camprubí and Philipp Lehmann, eds., "Experiencing the Global Environment," *Studies in History and Philosophy of Science Part A* 70 (2018).

38. On the global history of the use of eucalyptus trees to fight desertification, see Brett M. Bennett, "A Global History of Australian Trees," *Journal of the History of Biology* 44, no. 1 (February 1, 2011): 125–45; Brett M. Bennett, "The El Dorado of Forestry: The Eucalyptus in India, South Africa, and Thailand, 1850–2000," *International Review of Social History* 55, suppl. S18 (2010): 27–50. On some arguments for climate change through afforestation, see Oskar Lenz, *Timbuktu: Reise durch Marokko, die Sahara und den Sudan, ausgeführt im Auftrage der Afrikanischen Gesellschaft in Deutschland in den Jahren 1879 und 1880* (Leipzig: Brockhaus, 1884), 2:359–73; Louis Carton, "Climatologie et agriculture de l'Afrique ancienne," *Bulletin de l'Académie d'Hippone* 27 (1894): 1–45; Louis Carton, "Note sur la diminution des pluies en Afrique," *Revue tunisienne* 3 (1896): 87–94; Aleksandr Ivanovich Woeikof, "Der Einfluss der Wälder auf das Klima," *Petermanns Mitteilungen aus Justus Perthes' Geographischer Anstalt* 31, no. 3 (1885): 81–87; Eckardt, *Das Klimaproblem der geologischen Vergangenheit und historischen Gegenwart*, 128–29.

39. Franz von Czerny, *Die Veränderlichkeit des Klimas und ihre Ursachen* (Vienna: A. Hartleben, 1881), 4–5.

40. Brückner, *In wie weit ist das heutige Klima konstant?* On desiccation research in Central Asia, see David Moon, "Agriculture and the Environment on the Steppes in the Nineteenth Century," in *Peopling the Russian Periphery: Borderland Colonization in Eurasian History*, ed. Nicholas B. Breyfogle, Abby M. Schrader, and Willard Sunderland (London: Routledge, 2007), 81–105; David Moon, "The Debate over Climate Change in the Steppe Region in Nineteenth-Century Russia," *Russian Review* 69, no. 2 (April 1, 2010): 251–75; David Moon, *The Plough That Broke the Steppes: Agriculture and Environment on Russia's Grasslands, 1700–1914* (Oxford: Oxford University Press, 2013); Philippe Forêt, "Climate Change: A Challenge to the Geographers of Colonial Asia," *RFIEA Perspectives*, no. 9 (2013): 21–23; Deborah R. Coen, "Imperial Climatographies from Tyrol to Turkestan," *Osiris* 26 (2011): 45–65.

41. Prince Kropotkin, "The Desiccation of Eur-Asia," *Geographical Journal* 23, no. 6 (June 1904): 722–34. For some of the more prominent reactions to Kropotkin's article, see Prince Kropotkin, John Walter Gregory, and Edmond Cotter, "Correspondence: On the

Desiccation of Eurasia and Some General Aspects of Desiccation," *Geographical Journal* 43, no. 4 (1914): 451–59; Lev Berg, "Ist Zentral-Asien im Austrocknen begriffen?," *Geographische Zeitschrift* 13 (1907): 568–79.

42. Nils Ekholm, "On the Variations of the Climate of the Geological and Historical Past and Their Causes," *Quarterly Journal of the Royal Meteorological Society* 27, no. 117 (1901): 1–62; Johan Gunnar Andersson, "Das spätquartäre Klima: Eine zusammenfassende Übersicht über die in dieser Arbeit vorliegenden Berichte," in *Die Veränderungen des Klimas seit dem Maximum der letzten Eiszeit, eine Sammlung von Berichten unter Mitwirkung von Fachgenossen in verschiedenen Ländern*, ed. Johan Gunnar Andersson (Stockholm: Generalstabens Litografiska Anstalt, 1910), lvi.

43. Joseph Partsch, "Über den Nachweis einer Klimaänderung der Mittelmeerländer in geschichtlicher Zeit," in *Verhandlungen des VIII. deutschen Geographentages* (Berlin: W. Pormetter, 1889), 116–25. Around the same time, Julius Hann voiced similar doubts; see Nico Stehr and Hans von Storch, "Klimawandel, Klimapolitik und Gesellschaft," in *Eduard Brückner—Die Geschichte unseres Klimas: Klimaschwankungen und Klimafolgen*, ed. Nico Stehr and Hans von Storch (Vienna: Zentralanstalt für Meteorologie und Geodynamik, 2008), 16. The issue of focusing on weather extremes remains an issue in today's climate change debate; see Nico Stehr and Hans von Storch, "The Social Construction of Climate and Climate Change," *Climate Research* 5, no. 2 (1995): 103.

44. Cushman, "Humboldtian Science"; Partsch, "Über den Nachweis einer Klimaänderung der Mittelmeerländer," 122–24; Hermann Vogelstein, *Die Landwirtschaft in Palästina zur Zeit der Mišnâh. 1. Teil: Der Getreidebau* (Berlin: Mayer & Müller, 1894). The Mishnah (in Hebrew מִשְׁנָה), redacted in the second to third century CE, is the first major written record of Jewish oral traditions that date from the sixth century BCE to the first century CE; Andrew Ellicott Douglass, *Climatic Cycles and Tree-Growth: A Study of the Annual Rings of Trees in Relation to Climate and Solar Activity*, 3 vols., vol. 1 (Washington, DC: Carnegie Institution, 1919).

45. Julius Hann's influential work on climatology was central in forming this growing consensus; see Deborah R. Coen, "Climate and Circulation in Imperial Austria," *Journal of Modern History* 82, no. 4 (December 1, 2010): 846; Deborah R. Coen, *Climate in Motion: Science, Empire, and the Problem of Scale* (Chicago: University of Chicago Press, 2018), 160–217; Sluyter, *Colonialism and Landscape*, 223–27.

46. For a short overview of the competing theories, see Spencer Weart, *The Discovery of Global Warming* (Cambridge, MA: Harvard University Press, 2003), 1–18. See also Nico Stehr and Hans von Storch, *Klima, Wetter, Mensch* (Opladen: Barbara Budrich, 2010), 77–79; Robert P. Beckinsale, Richard J. Chorley, and J. Dunn, eds., *The History of the Study of Landforms, or, the Development of Geomorphology*, vol. 3 (London: Routledge, 1991), 48–56—in the book, Beckinsale and Chorley list the following predominant terrestrial theories of climate change mechanisms that existed alongside astronomical and solar theories between 1890 and 1920: (1) continental drift; (2) changes in land and sea distribution; (3) changes in land elevation; (4) changes in ocean circulation; (5) a long-term decrease in atmospheric carbon dioxide; (6) an increase in volcanic dust (54–55). For an overview of astronomical theories, see André Berger, "A Brief History of the Astronomical Theories of Paleoclimates," in *Climate Change*, ed. André Berger, Fedor Mesinger, and Djordje Sijacki (Vienna: Springer, 2012), 107–29.

47. Eduard Brückner, *Klimaschwankungen seit 1700, nebst Bemerkungen über die Klimaschwankungen der Diluvialzeit*, Geographische Abhandlungen 4 (Vienna: Ed. Hölzel, 1890),

240–43. What made it especially difficult for Brückner was that his proposed climatic cycles did not conform to the much shorter periodicity of sunspots, which had become a much-debated phenomenon in meteorological circles since Heinrich Schwabe's studies in the middle of the nineteenth century; see Samuel Heinrich Schwabe, "Sonnenbeobachtungen im Jahre 1843," *Astronomische Nachrichten* 21 (February 1, 1844): 233; see also Michael Bean, "Heinrich Samuel Schwabe, 1789–1875," *Journal of the British Astronomical Association* 85 (1975): 532–33; Douglas V. Hoyt and Kenneth H. Schatten, *The Role of the Sun in Climate Change* (New York: Oxford University Press, 1997); Anderson, *Predicting the Weather*, 264ff.; Karl Hufbauer, *Exploring the Sun: Solar Science since Galileo* (Baltimore: Johns Hopkins University Press, 1991), 42–80. In 1901, the importance of sunspots in meteorological and climatological thought was once again boosted by William Lockyer, who postulated a second-order sunspot cycle of about thirty-five years. This corresponded well with Brückner's climate cycles, but the doubts about the effect of sunspots were never entirely quashed. For Lockyer's hypotheses, see William J. S. Lockyer, "The Solar Activity 1833–1900," *Proceedings of the Royal Society of London* 68 (January 1, 1901): 285–300. Charles Abbot continued the research into the effects of sunspots and the variability of the solar constant on the weather and climate into the 1960s; see Charles Greeley Abbot and F. E. Fowle, "Volcanoes and Climate," *Smithsonian Miscellaneous Collections* 60, no. 29 (1913): 1–24; Charles Greeley Abbot, "Precipitation in Five Continents," *Smithsonian Miscellaneous Collections* 151, no. 5 (1967): 1–30; see also David DeVorkin, "Defending a Dream: Charles Greeley Abbot's Years at the Smithsonian," *Journal for the History of Astronomy* 21 (1990): 121–36.

48. Grove and Damodaran, "Imperialism, Intellectual Networks, and Environmental Change," 4348; W. F. Hume, "Climatic Changes in Egypt during Post-glacial Times," in *Die Veränderungen Des Klimas Seit Dem Maximum Der Letzten Eiszeit*, ed. Johan Gunnar Andersson (Stockholm: Generalstabens Litografiska Anstalt, 1910), 421–24; Henry Hubert, "Le dessèchement progressif en Afrique occidentale," *Bulletin du Comité d'études historiques et scientifiques de l'Afrique occidentale française* 3 (1920): 401–67.

49. Leiter, *Die Frage der Klimaänderung während geschichtlicher Zeit in Nordafrika*, 142; see also Alfred Philippson, *Das Mittelmeergebiet: seine geographische und kulturelle Eigenart*, 4th ed. (B. G. Teubner, 1922), 123–31; John Walter Gregory, "Is the Earth Drying Up?," *Geographical Journal* 43, no. 2 and 3 (1914): 307.

50. Today, it is clear that Theobald Fischer and his followers overestimated the climatic changes that had happened over the past two to three millennia. The climate of North Africa had actually remained rather stable before the onset of anthropogenic global warming, whose effects would become measurable only in the twentieth century. And yet, the rock carvings and paintings that Barth had discovered did not lie. There had, in fact, been cattle, large mammals, and much more water in the North African interior at one point. The large climatic shifts that Fischer identified, however, had happened around 6,000 years earlier.

51. See John Imbrie and Katherine Palmer Imbrie, *Ice Ages: Solving the Mystery* (Cambridge, MA: Harvard University Press, 1986), 61–175. It was only after the acceptance of Milankovitch's theories that Brückner's research into climate cycles enjoyed a renaissance (which is still continuing today); see Wolfgang H. Berger, Jürgen Pätzold, and Gerold Wefer, "A Case for Climate Cycles: Orbit, Sun and Moon," in *Climate Development and History of the North Atlantic Realm*, ed. Gerold Wefer et al. (Berlin: Springer, 2002), 101–23. For a critique of the state of geographical

knowledge about North Africa at the turn of the twentieth century, see Theobald Fischer, "Aufgaben und Streitfragen der Länderkunde des Mittelmeergebiets," *Petermanns Mitteilungen aus Justus Perthes' Geographischer Anstalt* 50 (1904): 174–76. On the importance of the airplane in the exploration of the Sahara, see Lindemann, *Die Wüste*, 43. The Hungarian explorer László Almásy, on whom the "English Patient" was modeled, was particularly important in this technological exploration of the Sahara, using both cars and airplanes on his journeys. He also continued the discussion on environmental and climatic changes in the desert and discovered further rock drawings with Leo Forbenius; see László Almásy, *Schwimmer in Der Wüste: Auf Der Suche Nach Der Oase Zarzura*, ed. Raoul Schrott and Michael Farin (Innsbruck: Haymon, 1997), 126, 132–34, 197–200.

52. Stehr and Storch, "Klimawandel, Klimapolitik und Gesellschaft," 12; Nico Stehr, "The Ubiquity of Nature: Climate and Culture," *Journal of the History of the Behavioral Sciences* 32, no. 2 (1996): 151–59.

53. Ellsworth Huntington, *The Pulse of Asia: A Journey in Central Asia Illustrating the Geographical Basis of History* (Boston: Houghton Mifflin, 1907); Ellsworth Huntington, *Palestine and Its Transformation* (Boston: Houghton Mifflin, 1911); Ellsworth Huntington, "Changes of Climate and History," *American Historical Review* 18, no. 2 (January 1, 1913): 213–32; Ellsworth Huntington, *Civilization and Climate*, 3rd ed. (New Haven, CT: Yale University Press, 1924). For a biography of Huntington, see Geoffrey J. Martin, *Ellsworth Huntington: His Life and Thought* (Hamden: Archon Books, 1973). On Huntington's environmental and climatic determinism, see Kent McGregor, "Huntington and Lovelock: Climatic Determinism in the 20th Century," *Physical Geography* 25, no. 3 (January 1, 2004): 237–50. The German term *Geist* carries the same double connotation of the English "spirit" (either "esprit" or "alcoholic beverage"); see Joseph Partsch, *Palmyra, eine historisch-klimatische Studie* (Leipzig: B. G. Teubner, 1922).

54. Stehr and Storch refer to a "disappearance" of the climate change debate in the early twentieth century; see Stehr and Storch, *Klima, Wetter, Mensch*, 79. Despite a waning of interest, however, the debate on progressive and periodic climate changes continued. For three examples of articles from the 1920s in French, English, and German, see C. Rivière, "L'invariabilité du climat en Afrique du Nord depuis le début de la période historique," *Revue d'histoire naturelle appliquée* 1 (1920): 71–79, 136–40, 163–76, 197–202, 234–38, 263–67, 302–4; Ellsworth Huntington and Stephen Sargent Visher, *Climatic Changes: Their Nature and Causes* (New Haven, CT: Yale University Press, 1922); Paul Kessler, *Das Klima der jüngsten geologischen Zeiten und die Frage einer Klimaänderung in der Jetztzeit* (Stuttgart: Schweizerbart, 1923).

55. Grove and Damodaran, "Imperialism, Intellectual Networks, and Environmental Change," 4347.

56. Johannes Walther, *Das Gesetz der Wüstenbildung in Gegenwart und Vorzeit*, 2nd ed. (Leipzig: Quelle & Meyer, 1912), 99. On Walther, see Ilse Seibold, *Der Weg zur Biogeologie, Johannes Walther, 1860–1937: ein Forscherleben im Wandel der deutschen Universität* (Berlin: Springer-Verlag, 1992). On this particularly optimistic view of climate modification, see Czerny, *Die Veränderlichkeit des Klimas und ihre Ursachen*, 75.

57. Fischer, *Studien über das Klima der Mittelmeerländer*, 41.

58. See, e.g., Czerny, *Die Veränderlichkeit des Klimas und ihre Ursachen*, 74; Partsch, "Über den Nachweis einer Klimaänderung der Mittelmeerländer," 123–24; Brückner, *Klimaschwankungen seit 1700*, 12.

Chapter 2

1. Originally, Ibsen did not intend for *Peer Gynt* to be put on stage. He wrote the piece as a dramatic poem in 1867 and only some years later agreed for an abridged version to be produced for the theater, for which Edvard Grieg wrote the music.

2. Henrik Ibsen, *Peer Gynt: A Dramatic Poem*, trans. Charles Archer and William Archer (New York: Walter Scott, 1905), 154–55.

3. Jules Verne, *L'Iinvasion de la mer*, Voyages extraordinaires 54 (Paris: Hetzel, 1905); for an English translation, see Jules Verne, *Invasion of the Sea*, ed. Arthur B. Evans, trans. Edward Baxter (Middletown, CT: Wesleyan University Press, 2001).

4. Jean-Pierre Picot, *Le testament de Gabès: l'invasion de la mer (1905), ultime roman de Jules Verne* (Pessac: Presses universitaires de Bordeaux, 2004). Jules Verne had already featured Roudaire's project in his 1877 novel Hector Servadac (translated as *Off on a Comet* into English): Jules Verne, *Hector Servadac: voyages et aventures à travers le monde solaire* (Paris: Hetzel, 1878); Edwin E. Slosson, "Plans to Restore the City of Brass," *Science News-Letter* 14, no. 401 (December 15, 1928): 367.

5. François Élie Roudaire, "Une mer intérieure en Algérie," *Revue des deux Mondes* 44, no. 3 (May 15, 1874): 343.

6. Lucien Lanier, *L'Afrique. Choix de lectures de géographie*, 4th ed. (Paris: E. Belin, 1887), 146–49. Zaimeche reports that there were about five million hectares of forest in Algeria at the beginning of French colonization. This is only a little more than 2 percent of the 919,595 square miles of Algeria's surface area; see S. E. Zaimeche, "Change, the State and Deforestation: The Algerian Example," *Geographical Journal* 160, no. 1 (1994): 51; A. Mtimet, R. Attia, and H. Hamrouni, "Evaluating and Assessing Desertification in Arid and Semi-Arid Areas of Tunisia," in *Global Desertification: Do Humans Cause Deserts?*, ed. J. F. Reynolds and D. M. Stafford Smith (Berlin: Dahlem University Press, 2002), 198; Jean-François Troin et al., *Le Maghreb: hommes et espaces* (Paris: A. Colin, 1985), 21–22; René Arrus, *L'eau en Algérie de l'impérialisme au développement, 1830–1962* (Alger: Presses universitaires de Grenoble, 1985), 13; CGIAR CSI Consortium for Spatial Information, "Global Aridity and PET Database," https://cgiarcsi.community /2019/01/24/global-aridity-index-and-potential-evapotranspiration-climate-database-v3/; European Commission Joint Research Centre, "World Atlas of Desertification," https://wad.jrc .ec.europa.eu/patternsaridity.

7. For a history of North African resistance—both passive and active—to French colonization attempts in Algeria, see Julia A. Clancy-Smith, *Rebel and Saint: Muslim Notables, Populist Protest, Colonial Encounters: Algeria and Tunisia, 1800–1904* (Berkeley: University of California Press, 1994). On French colonial violence in the Sahara, see Benjamin Claude Brower, *A Desert Named Peace: The Violence of France's Empire in the Algerian Sahara, 1844–1902* (New York: Columbia University Press, 2009), 27–90. On the "tentative, inchoate, and often contradictory" environmental representations of the Sahara in French thought, see George R. Trumbull, "Body of Work: Water and the Reimagining of the Sahara in the Era of Decolonization," in *Environmental Imaginaries of the Middle East and North Africa*, ed. Diana K. Davis and Edmund Burke (Athens: Ohio University Press, 2011), 87–112. On the first French projects in colonial Algeria, see Arnaud Berthonnet, "La formation d'une culture économique et technique en Algérie (1830–1962): l'exemple des grandes infrastructures de génie civil," *French Colonial History* 9, no. 1 (2008): 39–41.

8. Berthonnet, "La formation d'une culture économique et technique en Algérie," 41–42; Arrus, *L'eau en Algérie*, 38–39.

9. John Ruedy, *Modern Algeria: The Origins and Development of a Nation*, 2nd ed. (Bloomington: Indiana University Press, 2005), 55–68; James McDougall, *A History of Algeria* (New York: Cambridge University Press, 2017), 49–85; Charles-André Julien, *Histoire de l'Algérie contemporaine, la conqête et les débuts de la colonisation, 1827–1871* (Alger: Casbah éditions, 2005); Arrus, *L'eau en Algérie*, 46, 63–64; Fabienne Fischer, *Alsaciens et Lorrains en Algérie: histoire d'une migration, 1830–1914* (Nice: Gandini, 1999), 63–114.

10. Dominique Tabutin, Jean-Noël Biraben, and Eric Vilquin, *L'histoire de la population de l'Afrique du Nord pendant le deuxième millénaire* (Louvain-la-Neuve: Université catholique de Louvain, Département des sciences de la population et du développement, 2002), 9; cf. Dorothy Good, "Notes on the Demography of Algeria," *Population Index* 27, no. 1 (January 1, 1961): 7; Ruedy, *Modern Algeria*, 80–98; McDougall, *History of Algeria*, 86–129. On the legal history of French Algeria, see Allan Christelow, *Muslim Law Courts and the French Colonial State in Algeria* (Princeton, NJ: Princeton University Press, 1985).

11. Diana K. Davis, *Resurrecting the Granary of Rome: Environmental History and French Colonial Expansion in North Africa* (Athens: Ohio University Press, 2007); Caroline Ford, "Reforestation, Landscape Conservation, and the Anxieties of Empire in French Colonial Algeria," *American Historical Review* 113, no. 2 (April 2008): 341–62; Caroline Ford, *Natural Interests: The Contest over Environment in Modern France* (Cambridge, MA: Harvard University Press, 2016), 138–63; Andrea E. Duffy, "Civilizing through Cork: Conservationism and la Mission Civilisatrice in French Colonial Algeria," *Environmental History* 23, no. 2 (April 1, 2018): 270–92. The increased rate of deforestation in the nineteenth century reflected a worldwide trend; see J. F. Richards and Richard P. Tucker, eds., *Global Deforestation and the Nineteenth-Century World Economy* (Durham, NC: Duke University Press, 1983), xi–xii; James Fairhead and Melissa Leach, *Misreading the African Landscape: Society and Ecology in a Forest-Savanna Mosaic* (Cambridge: Cambridge University Press, 1996). While the climate of North Africa and the Mediterranean region in general did not experience a steady decline, it also was not stable, experiencing alternating periods of temporary stability and change, which can be traced with paleoclimatic methods; see, e.g., Michael McCormick et al., "Climate Change during and after the Roman Empire: Reconstructing the Past from Scientific and Historical Evidence," *Journal of Interdisciplinary History* 43, no. 2 (October 2012): 169–220; Brent D. Shaw, "Climate, Environment, and History: The Case of Roman North Africa," in *Climate and History: Studies in Past Climates and Their Impact on Man*, ed. T. M. L. Wigley, M. J. Ingram, and G. Farmer (Cambridge: Cambridge University Press, 1981), 379–403; Kyle Harper and Michael McCormick, "Reconstructing the Roman Climate," in *The Science of Roman History: Biology, Climate, and the Future of the Past*, ed. Walter Scheidel (Princeton, NJ: Princeton University Press, 2018), 11–52. On French colonial forestry politics in Algeria, see Andrea E. Duffy, *Nomad's Land: Pastoralism and French Environmental Policy in the Nineteenth-Century Mediterranean World* (Lincoln: University of Nebraska Press, 2019); Henry Sivak, "Legal Geographies of Catastrophe: Forests, Fires, and Property in Colonial Algeria, *Geographical Review* 103, no. 4 (October 2013): 556–74.

12. ANOM GGA P/59: Ligue du Reboisement, "Appel aux Algériens," 1883. For more information on the Reforestation League, see Davis, *Resurrecting the Granary of Rome*, 108–23; Louis Carton, "Climatologie et agriculture de l'Afrique ancienne," *Bulletin de l'Académie d'Hippone* 27

(1894): 1–45; Louis Carton, "Note sur la diminution des pluies en Afrique," *Revue tunisienne* 3 (1896): 87–94.

13. Brower, *A Desert Named Peace*, 199–221. For detailed accounts of Roudaire's project in the context of domestic and colonial politics in France, see René Létolle and Hocine Bendjoudi, *Histoires d'une mer au Sahara: utopies et politiques* (Paris: Harmattan, 1997); Jean-Louis Marçot, *Une mer au Sahara: mirages de la colonisation, Algérie et Tunisie, 1869–1887* (Paris: Différence, 2003). For a short account of Roudaire's project in the context of French Sahara exploration in the nineteenth century, see Numa Broc, "Les Français face à l'inconnue saharienne: géographes, explorateurs, ingénieurs (1830–1881)," *Annales de Géographie* 96, no. 535 (1987): 325–29.

14. From the Arabic شطّ (*šaṭṭ*), meaning bank or embankment.

15. François Élie Roudaire, *La mer intérieure africaine* (Paris: Imprimerie de la Société anonyme de publications périodiques, 1883), 3; Roudaire, "Une mer intérieure en Algérie," 325; Roudaire, *La mer intérieure africaine*, 4–7; AAS "Commission des chotts": Ferdinand de Lesseps, "M. Roudaire. Rapport sur la mission des Schotts" (manuscript), December 11, 1876.

16. The *Periplus of the Mediterranean*, published under Scylax's name, was probably not written by the Greek traveler and geographer of the sixth century BC. It represents a collection of travel accounts by various travelers from the fourth century BC.

17. Roudaire, "Une mer intérieure en Algérie," 327–34; Roudaire, *La mer intérieure africaine*, 14–22.

18. Thomas Shaw, *Travels, or Observations Relating to Several Parts of Barbary and the Levant*, 2nd ed. (London: A. Millar and W. Sandby, 1757), 126, 148; James Rennell, *The Geographical System of Herodotus Examined and Explained by a Comparison with Those of Other Ancient Authors and with Modern Geography* (London: W. Bulmer, 1800), 659–67; Konrad Mannert, *Geographie der Griechen und Römer*, 10 vols. (Nuremberg: E.C. Grattenauer, 1799). For a heavily edited abridged French version that was used in the debates over Roudaire's plans, see Konrad Mannert, *Géographie ancienne des États barbaresques d'après l'allemand de Mannert*, ed. Ludwig Marcus and F. Duesberg (Paris: Librairie encyclopédique de Roret, 1842). Roudaire's emphasis on classical sources and the re-creation of classical conditions is a compelling example of the general trend toward combining modern scientific methods and classical geographies to "[give] form and credibility to grand visions"; see Veronica della Dora, "Geo-Strategy and the Persistence of Antiquity: Surveying Mythical Hydrographies in the Eastern Mediterranean, 1784–1869," *Journal of Historical Geography* 33, no. 3 (July 2007): 516. On the importance of classical—and in particular Roman—models in the French colonization of Algeria and Tunisia, see Jacques Frémeaux, "Souvenirs de Rome et présence française au Maghreb: Essai d'investigation," in *Connaissances du Maghreb. Sciences sociales et colonisation*, ed. Jean Claude Vatin (Paris: CNRS, 1984), 29–46; Patricia M. E. Lorcin, "Rome and France in Africa: Recovering Colonial Algeria's Latin Past," *French Historical Studies* 25, no. 2 (April 1, 2002): 295–329; Nabila Oulebsir, *Les usages du patrimoine: monuments, musées et politique coloniale en Algérie, 1830–1930* (Paris: Maison des sciences de l'homme, 2004), 159–62. On the environmental dimension of the Roman model for French colonization in North Africa, see Diana K. Davis, "Restoring Roman Nature: French Identity and North African Environmental History," in *Environmental Imaginaries of the Middle East and North Africa*, ed. Diana K. Davis and Edmund Burke (Athens: Ohio University Press, 2011), 60–86.

19. Charles Martins, "Le Sahara: Souvenirs d'un voyage d'hiver," *Revue des deux Mondes* 34, no. 4 (July 15, 1864): 314. Extracts of Martins' long article were published as Charles Martins,

Tableau physique du Sahara oriental de la province de Constantine. Souvenirs d'un voyage exécuté pendant l'hiver de 1863 dans l'Oued-Riz et dans l'Oued-Souf (Paris: J. Claye, 1864); Henri Duveyrier, *Les Touareg du Nord* (Paris: Challamel, 1864), 42; Michael Heffernan, "The Limits of Utopia: Henri Duveyrier and the Exploration of the Sahara in the Nineteenth Century," *Geographical Journal* 155, no. 3 (November 1989): 344; see the debate over governmental aid for Duveyrier's expeditions in ANOM GGA 53S/1; Georges Lavigne, "Le percement de Gabès," *Revue moderne* 55 (1869): 322–35.

20. Dubocq, "Mémoire sur la constitution géologique des Zibân et de l'Ouad R'ir au point de vue des eaux artésiennes de cette portion du Sahara," *Annales des mines* 5, no. 2 (1852): 249–330; Roudaire, "Une mer intérieure en Algérie," 326; Henry Chotard, *La mer intérieure du Sahara* (Clermont-Ferrand: G. Mont-Louis, 1879), 11; Agnes Murphy, *The Ideology of French Imperialism, 1871–1881* (Washington, DC: Catholic University of America Press, 1948), 71–72.

21. AdS "Commission des chotts": François Roudaire, *Sur les travaux de la mission chargée d'étudier le projet de Mer intérieure en Algérie; communication faite a la Société de géographie le 14 juillet 1875* (Paris: E. Martinet, 1875), 4–9, 11; see also François Élie Roudaire, "Sur les travaux de la mission chargée d'étudier le projet de mer intérieure en Algérie," *Comptes rendus hebdomadaires des séances de l'Académie des Sciences* 80 (1875): 1593–96. The results of Roudaire's first trip to the chotts were first published in Henri Duveyrier, "Premier rapport sur la mission des chotts du Sahara de Constantine," *Bulletin de la Société de géographie de Paris* 9 (1875): 482–503; François Élie Roudaire, "La mission des chotts du Sahara de Constantine," *Bulletin de la Société de géographie de Paris* 10 (1875): 113–25; François Élie Roudaire, "Rapport sur les opérations de la mission des chotts," *Bulletin de la Société de géographie de Paris* 10 (1875): 574–86.

22. François Élie Roudaire, *Rapport à M. le ministre de l'instruction publique sur la mission des chotts. Études relatives au projet de mer intérieure* (Paris: Imprimerie nationale, 1877); originally, this report appeared as François Élie Roudaire, "Rapport à M. le ministre de l'instruction publique sur la mission des chotts. Études relatives au projet de mer intérieure," *Archives des missions scientifiques et littéraires* 4 (1877): 157–271; Michael Heffernan, "Bringing the Desert to Bloom: French Ambitions in the Sahara Desert during the Late Nineteenth Century—The Strange Case of 'La Mer Intérieure,'" in *Water, Engineering, and Landscape: Water Control and Landscape Transformation in the Modern Period*, ed. Denis E. Cosgrove and Geoffrey E. Petts (London: Belhaven Press, 1990), 102; Hippolyte Gautier and Adrien Desprez, *Les curiosités de l'Exposition de 1878: guide du visiteur* (Paris: C. Delagrave, 1878), 134; C. Delvaille, *Notes d'un visiteur sur l'Exposition universelle de 1878* (Paris: C. Delagrave, 1879), 119–20; Edmond Villetard, "A travers l'Exposition universelle," *Le Correspondant* 111 (1878): 919.

23. François Élie Roudaire, "Rapport à M. le ministre de l'instruction publique sur la dernière expédition des chotts. Complément des études relatives au projet de mer intérieure," *Archives des missions scientifiques et littéraires* 7 (1881): 231–413; see also Marçot, *Une mer au Sahara*, 342–51.

24. Roudaire, *Rapport sur la mission des chotts*, 86; Roudaire, *La mer intérieure africaine*, 94–96; Roudaire, "Une mer intérieure en Algérie," 349.

25. Roudaire, "Une mer intérieure en Algérie," 342.

26. AAS "Commission des chotts": Roudaire to the Academy of Sciences in Paris, March 9, 1877; in the letter, Roudaire included citations from Antoine César Becquerel and Edmond Becquerel, *Éléments de physique terrestre et de météorologie* (Paris: Firmin Didot, 1847), 104–5,

170–71. Through increases in low stratus clouds and fog, and the change of precipitation patterns, reservoirs do indeed have an effect on local climates; see R. M. Baxter, "Environmental Effects of Dams and Impoundments," *Annual Review of Ecology and Systematics* 8 (January 1, 1977): 255–83; C. J. Vörösmarty et al., "Drainage Basins, River Systems, and Anthropogenic Change: The Chinese Example," in *Asian Change in the Context of Global Climate Change*, ed. James Galloway and Jerry M. Melillo (Cambridge: Cambridge University Press, 1998), 210–44. What exactly the climatic effect will be in each case, however, is extremely difficult to predict; on the complexity of climate models and the difficulties in predicting the effects of anthropogenic climate change on the local level, see Roger A. Pielke and William R. Cotton, *Human Impacts on Weather and Climate*, 2nd ed. (Cambridge: Cambridge University Press, 2007), 102–50, 243–54; Lanier, *L'Afrique*, 339–40; Roudaire, *La mer intérieure africaine*, 92.

27. Ferdinand de Lesseps, "Communication sur les lacs amers de l'isthme de Suez," *Comptes rendus hebdomadaires des séances de l'Académie des Sciences* 78 (1874): 1740–48; Roudaire, *La mer intérieure africaine*, 26; Roudaire, "Une mer intérieure en Algérie," 348.

28. Roudaire, *La mer intérieure africaine*, 93; for Tyndall's findings, see John Tyndall, "On the Absorption and Radiation of Heat by Gases and Vapours and on the Physical Connexion of Radiation, Absorption and Conduction," *Philosophical Transactions of the Royal Society of London* 151 (1861): 1–36. For Roudaire's notes on Tyndall, see AAS "Commission des chotts": Roudaire, "Extraits de Tyndall" (manuscript), n.d.

29. Roudaire, *La mer intérieure africaine*, 23–25; Christopher L. Hill, *National History and the World of Nations: Capital, State, and the Rhetoric of History in Japan, France, and the United States* (Durham, NC: Duke University Press, 2008), 147–48.

30. Sara B. Pritchard, "From Hydroimperialism to Hydrocapitalism: 'French' Hydraulics in France, North Africa, and Beyond," *Social Studies of Science* 42, no. 4 (August 1, 2012): 591–615.

Chapter 3

1. August Petermann, "Gerhard Rohlfs' neues afrikanisches Forschungs-Unternehmen," *Mittheilungen aus Justus Perthes' geographischer Anstalt* 24 (1878): 20–21.

2. On the important connection between water, environmental transformation, and political power, see Sara B. Pritchard, "From Hydroimperialism to Hydrocapitalism: 'French' Hydraulics in France, North Africa, and Beyond," *Social Studies of Science* 42, no. 4 (August 1, 2012): 591–615.

3. Benjamin Claude Brower, *A Desert Named Peace: The Violence of France's Empire in the Algerian Sahara, 1844–1902* (New York: Columbia University Press, 2009), 54–64, 75–89; Eugène Daumas, *Le Sahara algérien: études géographiques, statistiques et historiques sur la région au sud des établissements français en Algérie* (Paris: Fortin, Masson & Langlois et Leclercq, 1845).

4. Rouire somewhat puzzlingly located the "split" in 1883, when the debate was already more or less over; see Alphonse Marie Ferdinand Rouire, *La découverte du bassin hydrographique de la Tunisie centrale et l'emplacement de l'ancien lac Triton (ancienne mer intérieure d'Afrique)* (Paris: Challamel, 1887), xii; a more appropriate dating of the academic disagreement would have to refer to 1874–75, when Roudaire's early publications on the Sahara Sea were critically examined for the first time.

5. AMAE 49 MD/4: François Philippe Voisin-Bey, "Examen technique du projet de mer intérieure de M. le Commandant Roudaire: Rapport à M. le Ministre de l'Instruction Publique

sur la dernière expedition des chotts" (manuscript), August 1881, 22–59; AAS "Commission des chotts": Yvon Villarceau, "Rapports sur les travaux géodésiques et topographiques exécutés en Algérie par M. Roudaire, séance du 7 mai 1877"; Yvon Villarceau, "Rapports sur les travaux géodesiques et togopgraphiques, exécutés en Algérie, par M. Roudaire," *Comptes rendus hebdomadaires des séances de l'Académie des Sciences* 84 (1877): 1002–13; John Ball, "Problems of the Libyan Desert," *Geographical Journal* 70 (1927): 24.

6. Edmond Fuchs, "Note sur l'isthme de Gabès et l'extrémité orientale de la dépression Saharienne," *Comptes rendus de l'Académie des Sciences* 79 (1874): 352–55—three years later, Fuchs published an updated longer version of the article as: Edmond Fuchs, "Note sur l'isthme de Ghabès et l'extrémité orientale de la dépression saharienne," *Bulletin de la Société de géographie de Paris* 14 (September 1877): 248–76; Henry Chotard, *La mer intérieure du Sahara* (Clermont-Ferrand: G. Mont-Louis, 1879), 15; Ernest Cosson, "Note sur le projet d'établissement d'une mer intérieure en Algérie," *Comptes rendus hebdomadaires des séances de l'Académie des Sciences* 79 (1874): 435–42; Karl Alfred von Zittel, "Das Saharameer," *Das Ausland* 56 (1883): 524–28; Erwin von Bary, "Reisebriefe aus Nord-Afrika," *Zeitschrift der Gesellschaft für Erdkunde zu Berlin.* 12 (1877): 196–98.

7. C. R. Pennell, *Morocco since 1830: A History* (New York: New York University Press, 2000), 354; Karl Alfred von Zittel, *Die Sahara: Ihre physische und geologische Beschaffenheit* (Kassel: Theodor Fischer, 1883), 40; AAS "Commission des chotts": Gabriel Auguste Daubrée, "Note à classer dans le dossier de la Commission des Chotts" (manuscript), n.d. (ca. 1878).

8. ANOM GGA P/59: Ligue du Reboisement, La forêt: conseils aux indigènes (Algiers: P. Fontana, 1883), 2; see also Diana K. Davis, *Resurrecting the Granary of Rome: Environmental History and French Colonial Expansion in North Africa* (Athens: Ohio University Press, 2007), 45–130; Pritchard, "From Hydroimperialism to Hydrocapitalism," 596; John Ruedy, *Modern Algeria: The Origins and Development of a Nation*, 2nd ed. (Bloomington: Indiana University Press, 2005), 91.

9. François Trottier, *Rôle de l'Eucalyptus en Algérie au point de vue des besoins locaux de l'exportation et du développement de la population* (Algiers: Aillaud, 1876), 89–93; see also François Trottier, *Boisement dans le désert et colonisation* (Algiers: F. Paysant, 1869). For more on Trottier, see Davis, *Resurrecting the Granary of Rome*, 104–8. On the global history of eucalyptus, see Brett M. Bennett, "A Global History of Australian Trees," *Journal of the History of Biology* 44, no. 1 (February 1, 2011): 125–45.

10. Emile Louis Bertherand, *Hygiène publique: Malaria et forêts en Algérie d'après une enquête de la Société climatologique d'Alger* (Algiers: P. Fontana, 1882); Antoine François Thévenet, "Le service météorologique algérien," in *Association française pour l'avancement des sciences. Compte rendu de la 17me session, Oran 1888*, 2 vols. (Paris: Imprimerie de Chaix, 1888), 2:233–37. By 1900, the service oversaw forty-four meteorological stations in Algeria, some of which were situated in the southernmost permanently settled regions of the country; see Charles de Galland, *Renseignements sur l'Algérie: les petits cahiers algériens* (Algiers: A. Jourdan, 1900), 88; *Rapport sur les observatoires astronomiques de province* (Paris: Imprimerie national, 1902), 12–13. On the difficult beginnings of the service and the continuing problems into the twentieth century, see Michael A. Osborne, *Nature, the Exotic, and the Science of French Colonialism* (Bloomington: Indiana University Press, 1994), 150–51, 169–70; A. Angot, "Le régime des vents et l'évaporation dans la région des chotts algériens," *Comptes rendus hebdomadaires des séances de l'Académie des Sciences* 85 (1877): 396–99.

In 1896, the first comprehensive study of the Algerian climate showed that the prevalent winds in the region of the chotts (measured at the meteorological stations on Djerba and in Biskra) blew from the west in the winter and from the east in the summer. In both cases, the moisture from the projected inland sea would have been lost in the sea and in the sand desert, respectively. See Antoine François Thévenet, *Essai de climatologie algérienne* (Algiers: Giralt, 1896), 30–31.

11. Oskar Lenz, "Kurzer Bericht über meine Reise von Tanger nach Timbuktu und Senegambien," *Zeitschrift der Gesellschaft für Erdkunde zu Berlin* 16 (1881): 291–92; Zittel, "Das Saharameer," 528.

12. Martin J. S. Rudwick, *Worlds Before Adam: The Reconstruction of Geohistory in the Age of Reform* (Chicago: University of Chicago Press, 2008). Desor chronicled Agassiz's research in the Alps; see Eduard Desor, *Excursions et séjours dans les glaciers et les hautes régions des Alpes de M. Agassiz et ses compagnons de voyage* (Neuchâtel: Kissling, 1844). A few years later, Agassiz and Desor had a public falling out over the original authorship of zoological research, which quickly turned personal with allegations of debt evasion and home wrecking; see the documents chronicling the mediation attempts in HLHU MS Am 1419, Series II—Agassiz v. Desor. See also Christoph Irmscher, *Louis Agassiz: Creator of American Science* (Boston: Houghton Mifflin Harcourt, 2013), 98–102; Eduard Desor, *Aus Sahara und Atlas: Vier Briefe an J. Liebig* (Wiesbaden: C. W. Kreidel, 1865), 42, 53.

13. Georges Lavigne, "Le percement de Gabès," *Revue moderne* 55 (1869): 334.

14. Ferdinand de Lesseps, "La mer intérieure de Gabès," *Revue scientifique de la France et de l'étranger* 3 (April 1883): 496; Ferdinand de Lesseps, "Observations au sujet de l'établissement d'une mer intérieure en Algérie," *Comptes rendus hebdomadaires des séances de l'Académie des Sciences* 79 (1874): 88; Auguste Pomel, "Algérie. Nouvelle exploration de M. Roudaire," *Revue géographique internationale* 3, no. 32 (June 1878): 179; *Commission supérieure pour l'examen du projet de mer intérieure dans le sud de l'Algérie et de la Tunisie* (Paris: Imprimerie nationale, 1882), 411–18.

15. On the theory of a flooded Sahara during the Quaternary Period, see Arnold Escher von der Linth, "Die Gegend von Zürich in der letzten Period der Vorwelt," in *Zwei geologische Vorträge gehalten im März 1852* (Zurich: Kiesling, 1852); see also Oswald Heer, *Arnold Escher von der Linth: Lebensbild eines Naturforschers* (Zurich: F. Schulthess, 1873), 322–25. This theory was later attacked and quite convincingly disproved by the German "father" of meteorology, Heinrich Dove; see Heinrich Wilhelm Dove, *Der schweizer Föhn: Nachtrag zu Eiseit, Föhn und Scirocco* (Berlin: D. Reimer, 1868).

16. See John Tresch, *The Romantic Machine: Utopian Science and Technology after Napoleon* (Chicago: University of Chicago Press, 2012).

17. AAS "Commission des chotts": Favé, "Rapport sur les travaux géodésiques et topographiques," n.d., 17–18.

18. Timothy Mitchell, *Rule of Experts: Egypt, Techno-Politics, Modernity* (Berkeley: University of California Press, 2002), 80–120; Ignatius Frederick Clarke, "Almanac of Anticipations: A Prospect of Probabilities, 1830–1890," *Futures* 16, no. 3 (June 1984): 323; Dirk van Laak, *Weisse Elefanten: Anspruch und Scheitern technischer Grossprojekte im 20. Jahrhundert* (Stuttgart: Deutsche Verlags-Anstalt, 1999), 24–25.

19. On the global connections and exchanges of water engineering technology, see Jessica B. Teisch, *Engineering Nature: Water, Development, and the Global Spread of American Environmental Expertise* (Chapel Hill: University of North Carolina Press, 2011). For an analysis of Saint-Simon's

thought, see Pierre Musso, *La religion du monde industriel: analyse de la pensée de Saint-Simon* (La Tour d'Aigues: Aube, 2006); Dominique Casajus, *Henri Duveyrier: un saint-simonien au désert* (Paris: Ibis Press, 2007). For a prosopography of Saint-Simonians in France, see Fritzie Prigohzy Manuel and Frank Edward Manuel, *Utopian Thought in the Western World* (Cambridge, MA: Belknap, 1979), 590–640. On Saint-Simonianism and engineering, see Antoine Picon, "French Engineers and Social Thought, 18–20th Centuries: An Archeology of Technocratic Ideals," *History and Technology* 23, no. 3 (September 1, 2007): 197–208; Antoine Picon, *Les Saint-Simoniens: raison, imaginaire et utopie* (Paris: Belin, 2002); Pamela M. Pilbeam, *Saint-Simonians in Nineteenth-Century France: From Free Love to Algeria* (Basingstoke: Palgrave Macmillan, 2014).

20. Prosper Enfantin, *Colonisation de l'Algérie* (Paris: P. Bertrand, 1843); François Élie Roudaire, "Une mer intérieure en Algérie," *Revue des deux Mondes* 44, no. 3 (May 15, 1874): 350; *Commission supérieure pour l'examen du projet de mer intérieure*, 166–72, 182–83.

21. René Létolle and Hocine Bendjoudi, *Histoires d'une mer au Sahara: utopies et politiques* (Paris: Harmattan, 1997), 92, 143; Emil Deckert, *Die Kolonialreiche und Kolonisationsobjekte der Gegenwart: kolonialpolitische und kolonialgeographische Skizzen* (Leipzig: P. Frohberg, 1884), 120.

22. Despite newly won rights, the French did not exercise complete sovereignty in Tunisia, and not everyone in France was happy with the stipulations that left some measure of autonomy to the Bey of Tunis; see Mary Dewhurst Lewis, *Divided Rule: Sovereignty and Empire in French Tunisia, 1881–1938* (Berkeley: University of California Press, 2014), 28–32; Chotard, *La mer intérieure du Sahara*, 16.

23. AAS "Commission des chotts": Idelphonse Favé, "Rapports sur les travaux géodésiques et topographiques exécutés en Algérie par M. Roudaire, séance du 21 mai 1877"; Watteville to Roudaire, March 9, 1879, in G. Dubost, *Le colonel Roudaire et son projet de mer saharienne* (Guéret: Société des sciences naturelles et archéologiques de la Creuse, 1998), 75; François Élie Roudaire, *La mer intérieure africaine* (Paris: Imprimerie de la Société anonyme de publications périodiques, 1883), 36–43; AMAE 49 MD/4: Roudaire to Foreign Minister, Paris, June 28, 1881.

24. Freycinet to Jules Grévy, May 27, 1882, cited in Roudaire, *La mer intérieure africaine*, 37–39; AMAE 49 MD/4: Charles de Freycinet, "Rapport au Président de la République française," *Journal officiel de la République française* 14, no. 116 (April 28, 1882), 2241–43; *Commission supérieure pour l'examen du projet de mer intérieure*, 13–15. The debates and findings of the Commission supérieure on Roudaire's project can be found in AMAE 49 MD/5–7.

25. For Voisin's detailed account of the Suez Canal construction, see François Philippe Voisin, *Le canal de Suez*, 6 vols. (Paris: Dunod, 1902); AMAE 49 MD/4: François Philippe Voisin, "Examen technique du projet de mer intérieure"; AMAE 49 MD/4: Chambrelent, "Note sur la remplissage de la mer intérieure" (manuscript), June 24, 1882; *Commission supérieure pour l'examen du projet de mer intérieure*, 536.

26. *Commission supérieure pour l'examen du projet de mer intérieure*, 545–46; Ernest Cosson, "Note sur le projet de création, en Algérie et en Tunisie, d'une mer dite intérieure," *Comptes rendus hebdomadaires des séances de l'Académie des Sciences* 96 (1883): 1191–96; James C. Scott, *Seeing Like a State: How Certain Schemes to Improve the Human Condition Have Failed*, Yale Agrarian Studies (New Haven, CT: Yale University Press, 1998), 3.

27. Lesseps, "La mer intérieure de Gabès"; for other notes and statements of support for Roudaire's project, see M. Gellerat, *Note sur la mer intérieure africaine ou Mer Roudaire* (Paris: P. Dubreuil, 1883).

28. Michael Heffernan, "Bringing the Desert to Bloom: French Ambitions in the Sahara Desert during the Late Nineteenth Century—The Strange Case of 'La Mer Intérieure,'" in *Water, Engineering, and Landscape: Water Control and Landscape Transformation in the Modern Period*, ed. Denis E. Cosgrove and Geoffrey E. Petts (London: Belhaven Press, 1990), 107; René Arrus, *L'eau en Algérie de l'impérialisme au développement, 1830–1962* (Alger: Presses universitaires de Grenoble, 1985), 51–52.

29. ANOM GGA 3E95: E. Jacob to the Governor-General of Algeria, May 15, 1872; Benjamin Milliot, "Le dessèchement du lac Fetzara," in *Association française pour l'avancement des sciences. Comptes-rendus de la 10e session, Alger 1881* (Paris, 1882), 802–7.

30. George R. Trumbull, *An Empire of Facts: Colonial Power, Cultural Knowledge, and Islam in Algeria, 1870–1914* (Cambridge: Cambridge University Press, 2009), 212–24; Marcel Cassou, *Le Transsaharien: L'échec sanglant des Missions Flatters, 1881* (Paris: L'Harmattan, 2004); Mike Heffernan, "Shifting Sands: The Trans-Saharan Railway," in *Engineering Earth*, ed. Stanley D. Brunn (Dordrecht: Springer, 2011), 617–26; Jean-Louis Marçot, *Une mer au Sahara: mirages de la colonisation, Algérie et Tunisie, 1869–1887* (Paris: Différence, 2003), 359–62; Michael Heffernan, "The Limits of Utopia: Henri Duveyrier and the Exploration of the Sahara in the Nineteenth Century," *Geographical Journal* 155, no. 3 (November 1989): 342–52; Henri Duveyrier, *Les Touareg du Nord* (Paris: Challamel, 1864), 317–454; Trumbull, *Empire of Facts*, 224–37.

31. Lucien Lanier, *L'Afrique. Choix de lectures de géographie*, 4th ed. (Paris: E. Belin, 1887), 340; see also Gustave-Ernest-Alfred Landas, *Port et oasis du bassin des chotts tunisiens: projet de M. le Commandant Landas* (Paris: Société anonyme de publications périodiques, 1886).

32. Mackenzie explicitly referred to de Lesseps and his views that a submersion of the Sahara would have only beneficial effects on Europe's climate; see Donald Mackenzie, *The Flooding of the Sahara: An Account of the Proposed Plan for Opening Central Africa to Commerce and Civilization from the North-West Coast* (London: S. Low, Marston, Searle, & Rivington, 1877), xii–xiii; François Élie Roudaire, *Rapport à M. le ministre de l'instruction publique sur la mission des chotts. Études relatives au projet de mer intérieure* (Paris: Imprimerie nationale, 1877), 100; Pennell, *Morocco since 1830*, 55–58. The official correspondence regarding Mackenzie's projects in North Africa is collected in NA under the signature FO 881/4670; Donald Mackenzie, "North-West African Expedition," *The Times*, December 23, 1876, sec. Letters to the Editor.

33. Arthur Cotton, *The Story of Cape Juby* (London: Waterlow & Sons, 1894), 49–51.

34. Donald Mackenzie, "The British Settlement at Cape Juby, North-West Africa," *Blackwood's Edinburgh Magazine* 146, no. 887 (1889): 412; see also Clarke, "Almanac of Anticipations," 322; "The Curiosity of the World," *The Times*, August 7, 1875, sec. Editorials; Mackenzie, *Flooding of the Sahara*, 7–8.

35. NA FO 881/4670: Pauncefote to Consul Dundas, October 2, 1878, 11; Donald Mackenzie, *A Report on the Condition of the Empire of Morocco* (London: British and Foreign Anti-Slavery Society, 1886), 6, 19–21; Pennell, *Morocco since 1830*, 85–88; François Zuccarelli and Philippe Decraene, *Grands sahariens: à la découverte du "désert des déserts"* (Paris: Denoël, 1994), 179–80.

36. John D. Champlin, "The Proposed Inland Sea in Algeria," *Popular Science Monthly*, April 1876; John T. Short, "The Flooding of the Sahara," *Scribner's Monthly*, July 1879. Publications of the time drew direct connections between plans in the United States and Roudaire's project; see, e.g., "Governor Fremont's Projected Sea," *New York Tribune*, April 12, 1879; see also Andrew F. Rolle, *John Charles Frémont: Character as Destiny* (Norman: University of Oklahoma

Press, 1991), 251; Paul Staudinger, "Die algerisch-tunesischen Schotts und die Frage der Bewässerung der Depressionen," *Geographische Zeitschrift* 1 (1895): 692–97; Chotard, *La mer intérieure du Sahara*, 17; Paul Bourde, *A travers l'Algérie: Souvenirs de l'excursion parlementaire* (Paris: G. Charpentier, 1880), 168–69.

37. See Narcisse Faucon, *Le livre d'or de l'Algérie* (Paris: Challamel, 1890), 237–42; ANOM FM AFRIQUE/XII/4: Victor Levasseur, "Canal de Tombouctou à la Mer" (manuscript), 1896. For one of the few newspaper articles on the Oued Righ project, see "The Fertilization of the Sahara," *Australasian Pastoralist's Review* 2, no. 11 (1893): 1029; ANOM FM AFRIQUE/XII/4: French Colonial Ministry, "Note pour la Direction des Affaires commerciales et de la Colonisation," March 7, 1896; G. A. Thompson, "A Plan for Converting the Sahara Desert into a Sea," *Scientific American* 107, no. 6 (1912): 114–25; "To Make Ocean of Sahara: French Savant Plans Canal to Flood Big African Desert," *Chicago Daily Tribune*, October 15, 1911; "Proposes to Turn Sahara into a Sea: French Engineer's Scheme," *New York Times*, October 15, 1911; see also James Rodger Fleming, *Fixing the Sky: The Checkered History of Weather and Climate Control* (New York: Columbia University Press, 2010), 201.

38. Ernest H. L. Schwarz, *The Kalahari or Thirstland Redemption* (Cape Town: T. M. Miller, 1920). Schwarz had first intimated his ideas in Ernest H. L. Schwarz, "The Desiccation of Africa: The Cause and the Remedy," *South African Journal of Science* 14 (1919): 139–78. See also Meredith McKittrick, "An Empire of Rivers: The Scheme to Flood the Kalahari, 1919–1945," *Journal of Southern African Studies* 41, no. 3 (May 4, 2015): 485–504.

39. See William Willcocks, *Egyptian Irrigation*, 3rd ed., vol. 2 (London: E. & F. N. Spon, 1913), 676–717; H. J. L. Beadnell, *The Topography and Geology of the Fayum Province of Egypt* (Cairo: National Printing Department, 1905), 16–24; Ball, "Problems of the Libyan Desert"; John Ball, "The Qattara Depression of the Libyan Desert and the Possibility of Its Utilization for Power-Production," *Geographical Journal* 82, no. 4 (1933): 289–314.

40. Edwin E. Slosson, "Plans to Restore the City of Brass," *Science News-Letter* 14, no. 401 (December 15, 1928): 365. Among other things, Slosson was the first editor of Science Service, whose goal was to communicate scientific and technological developments to the public; see Katherine Pandora, "Popular Science in National and Transnational Perspective: Suggestions from the American Context," *Isis* 100, no. 2 (2009): 357; Paul Borchardt, "Neue Beiträge zur alten Geographie Nordafrikas und zur Atlantisfrage," *Zeitschrift der Gesellschaft für Erdkunde zu Berlin* 62 (1927): 197–216; Paul Borchardt, "Platos Insel Atlantis: Versuch einer Erklärung," *Petermanns Mitteilungen aus Justus Perthes' Geographischer Anstalt* 73 (1927): 19–32; Paul Borchardt, "Eine kulturgeographische Studienreise nach Süd-Tunis 1928," *Petermanns Mitteilungen aus Justus Perthes' Geographischer Anstalt* 74 (1928): 162–65.

41. Slosson, "Plans to Restore the City of Brass," 368; George Chetwynd Griffith, *The Great Weather Syndicate* (London: George Bell and Sons, 1906).

Chapter 4

* This chapter is based in large part on the papers of Herman Sörgel, kept at the Deutsches Museum in Munich (ADM NL 92). The collection is substantial, if not complete: Sörgel burned part of his "Atlantropa Archive" during the Second World War in order to make the rest more easily transportable; see ADM NL 92/140: "Die Geschichte Atlantropas," *Atlantropa Mitteilungen* 22 (1949).

1. R. G. Johnson, "Climate Control Requires a Dam at the Strait of Gibraltar," *EOS, Transactions of the American Geophysical Union* 78, no. 27 (1997): 277–80; Jochem Marotzke and Alistair Adcroft, "Comment on 'Climate Control Requires a Dam at the Strait of Gibraltar,'" *EOS, Transactions of the American Geophysical Union* 78, no. 45 (1997): 507. Ideas about building a dam at Gibraltar are still around today; see, e.g., Jim Gower, "A Sea Surface Height Control Dam at the Strait of Gibraltar," *Natural Hazards* 78, no. 3 (September 1, 2015): 2109–20.

2. While Atlantropa has received little scholarly attention in the English-speaking world, two German monographs on Sörgel's scheme were published in 1998. For a detailed description of the different stages of the project and its place among early ideas of European cooperation, see Alexander Gall, *Das Atlantropa-Projekt: Die Geschichte einer gescheiterten Vision. Herman Sörgel und die Absenkung des Mittelmeers* (Frankfurt am Main: Campus, 1998); for an analysis of the project with an emphasis on its relation to developments in contemporary architecture, see Wolfgang Voigt, *Atlantropa: Weltbauen am Mittelmeer, ein Architektentraum der Moderne* (Hamburg: Dölling und Galitz, 1998); for one of the few English-language treatments of Atlantropa (if also written by a German-born author, who became an important space-flight advocate in the United States), see Willy Ley, *Engineers' Dreams* (New York: Viking, 1966), 138–56. Some of the research for this and the following chapter has been published as Philipp N. Lehmann, "Infinite Power to Change the World: Hydroelectricity and Engineered Climate Change in the Atlantropa Project," *American Historical Review*, 121, no. 1 (2016): 70–100.

3. Herman Sörgel, *Atlantropa* (Zurich: Fretz & Wasmuth, 1932), 139.

4. For more information on Oskar von Miller, most famous today as the founder of the Deutsches Museum in Munich, see Wilhelm Füssl, *Oskar von Miller 1855–1934: eine Biographie* (Munich: Beck, 2005). On the history of the Walchensee Power Plant, see Thomas Parke Hughes, *Networks of Power: Electrification in Western Society, 1880–1930* (Baltimore: Johns Hopkins University Press, 1983), 334–50; Reinhard Falter, "Achtzig Jahre 'Wasserkrieg': Das Walchenseekraftwerk," in *Von der Bittschrift zur Platzbesetzung: Konflikte um technische Großprojekte,* ed. Ulrich Linse et al. (Berlin: J.H.W. Dietz Nachfolger, 1988), 63–127.

5. Herman Sörgel, *Theorie der Baukunst,* 2 vols. (Munich: Piloty & Loehle, 1918).

6. Otto Jessen, *Südwest-Andalusien; Beiträge zur Entwicklungsgeschichte, Landschaftskunde und antiken Topographie Südspaniens, insbesondere zur Tartessosfrage,* Ergänzungsheft zu Petermanns geographischen Mitteilungen 186 (Gotha: J. Perthes, 1924); Otto Jessen, *Die Straße von Gibraltar* (Berlin: D. Reimer, 1927). This finding was first described in detail by W. B. Carpenter and generated some opposition, whose most eminent voice was Charles Lyell; see W. B. Carpenter, "On the Gibraltar Current, the Gulf Stream, and the General Oceanic Circulation," *Proceedings of the Royal Geographical Society of London* 15, no. 1 (1871): 54–91; William B. Carpenter, "Further Inquiries on Oceanic Circulation," *Proceedings of the Royal Geographical Society of London* 18, no. 4 (1873): 301–407.

7. Sörgel, *Atlantropa*; ADM NL 92/140: "Die Geschichte Atlantropas," *Atlantropa Mitteilungen* 14 (1948).

8. On the political dimension of Sörgel's plans and his ideas for a unified Europe, see Alexander Gall, "Atlantropa: A Technological Vision of a United Europe," in *Networking Europe: Transnational Infrastructures and the Shaping of Europe, 1850–2000,* ed. Erik van der Vleuten and Arne Kaijser (Sagamore Beach, MA: Science History Publications, 2006), 99–128; Sörgel,

Atlantropa, 24, 126; ADM NL 92/140: "Die Geschichte Atlantropas," *Atlantropa Mitteilungen* 14 (1948).

9. H. G. Wells, *The Outline of History: Being a Plain History of Life and Mankind* (New York: Garden City, 1920), 1091–94. Sörgel's second book-length publication on Atlantropa actually carried the name "The Three Great 'A'" in its title: Herman Sörgel, *Die drei großen "A": Großdeutschland und italienisches Imperium, die Pfeiler Atlantropas* (Munich: Piloty & Loehle, 1938); Richard Nicolaus Coudenhove-Kalergi, *Revolution durch Technik* (Vienna: Paneuropa Verlag, 1932).

10. In all likelihood, Sörgel chose the comparison to the Niagara Falls deliberately, as Germans had become fascinated with the large hydropower station at the falls; see David Blackbourn, *The Conquest of Nature: Water, Landscape, and the Making of Modern Germany* (New York: Norton, 2006), 217–18; ADM NL 92/194: Sörgel, "Die Welt ohne Kohle," *Sonntags-Zeitung*, October 4, 1936; Sörgel, *Die drei großen "A,"* 80. There is no room here to give the history of fossil fuel anxieties the treatment it deserves; instead, see Fredrik Albritton Jonsson, *Enlightenment's Frontier: The Scottish Highlands and the Origins of Environmentalism* (New Haven, CT: Yale University Press, 2013). For Jevons's most clearly articulated argument, see William Stanley Jevons, *The Coal Question: An Inquiry Concerning the Progress of the Nation, and the Probable Exhaustion of Our Coal-Mines*, 2nd ed. (London: Macmillan, 1866). See also Nuno Luis Madureira, "The Anxiety of Abundance: William Stanley Jevons and Coal Scarcity in the Nineteenth Century," *Environment and History* 18, no. 3 (2012): 395–421. Jevons also had some very interesting ideas about the interrelation between sunspots and economic cycles; see William Stanley Jevons, "Commercial Crises and Sun-Spots," *Nature* 19, no. 472 (1878): 33–37. On the history of scarcity, see Fredrick Albritton Jonsson, John Brewer, Neil Fromer, and Frank Trentmann, eds., *Scarcity in the Modern World* (London: Bloomsbury, 2019).

11. Lewis Mumford, for instance, imagined a future world based on hydropower and solar energy and had hopes for a "change from the inorganic to the organic, from the destructive to the conservative utilization of land and energy"; see Lewis Mumford, *The Culture of Cities* (New York: Harcourt, Brace, 1938), 326. On the history of the harnessing of wind energy, see Matthias Heymann, *Die Geschichte der Windenergienutzung: 1890–1990* (Frankfurt am Main: Campus, 1995); Hughes, *Networks of Power*, 264, 313ff.; Werner Bätzing, *Die Alpen: Geschichte und Zukunft einer europäischen Kulturlandschaft* (Munich: Beck, 2003), 192; Blackbourn, *Conquest of Nature*, 217–20.

12. See Pier Angelo Toninelli, "Energy and the Puzzle of Italy's Economic Growth," *Journal of Modern Italian Studies* 15, no. 1 (2010): fig. 3; James Sievert, *The Origins of Nature Conservation in Italy* (Bern: Peter Lang, 2000), 87–88; J. R McNeill, *Something New under the Sun: An Environmental History of the Twentieth-Century World* (New York: Norton, 2000), 173–77; Samuel S. Wyer, *Digest of the Transactions of First World Power Conference Held at London, England, June 30 to July 12, 1924* (London: Humphries & Company, 1925), 10.

13. See W. G. Jensen, "The Importance of Energy in the First and Second World Wars," *Historical Journal* 11, no. 3 (January 1, 1968): 538–54; Heinrich Wilhelm Voegtle, *Die Wasserkraftnutzung und die Bedeutung der deutschen Wasserturbinen-Industrie* (Heidenheim a.d. Brenz: C. F. Rees, 1927), 25–26.

14. Arthur Lichtenauer, *Die geographische Verbreitung der Wasserkräfte in Mitteleuropa* (Würzburg: Kabitzsch & Mönnich, 1926), 51–53.

15. Herman Sörgel, "Das Panropa-Projekt als städtebauliche Darstellungsstudie," *Baumeister* 29, no. 5 (1931): 216; Sörgel, *Atlantropa*, 78, 119.

16. On the Atlantropa-inspired novels, see ADM NL 92/26, 92/65, 92/140, 92/186; Gall, *Das Atlantropa-Projekt*, 151–65. For a number of surviving examples of these novels, see Georg Güntsche, *Panropa* (Cologne: Gilde-Verlag, 1930); Titus Taeschner, *Atlantropa* (Berlin: Goldmann, 1935); Walther Kegel, *Dämme im Mittelmeer* (Berlin: Buchwarte, 1937); Titus Taeschner, *Eurofrika, die Macht der Zukunft* (Berlin: Buchwarte, 1938). For a discussion of Güntsche's novel, see Menno Spiering, "Engineering Europe: The European Idea in Interbellum Literature, the Case of Panropa," in *Ideas of Europe since 1914: The Legacy of the First World War*, ed. Michael J. Wintle and Menno Spiering (Houndmills, Basingstoke: Palgrave Macmillan, 2002), 177–99; Sörgel, *Atlantropa*, 70.

17. ADM NL 92/140: "Die Geschichte Atlantropas," *Atlantropa Mitteilungen* 14 (1948); Sörgel, *Atlantropa*, 125; Herman Sörgel, *Mittelmeer-Senkung, Sahara-Bewässerung (Panropa-Projekt)* (Leipzig: J. M. Gebhardt, 1929), 30. Sörgel diagnosed an overpopulation crisis in Europe, but framed the issue as one of space and "carrying capacity" rather than of reproductive politics. Sörgel did not hold any antinatalist positions. In fact, he thought that Europe's only chance of survival in the face of the large and populous continents of Asia and America was an expansion of both the population and the territory. On German debates over national "carrying capacities" in the 1920s, see Ulrike Schaz and Susanne Heim, *Berechnung und Beschwörung: Überbevölkerung, Kritik einer Debatte* (Berlin: Verlag der Buchläden Schwarze Risse/Rote Strasse, 1996), 29–32. For an overview of fears of both overpopulation and depopulation in Europe and the rise of a pronatalist stance, see the introduction in Maria Sophia Quine, *Population Politics in Twentieth-Century Europe: Fascist Dictatorships and Liberal Democracies*, Historical Connections (London: Routledge, 1996), 1–16. For a contemporary example, see the work of the German demographer Friedrich Burgdörfer: Friedrich Burgdörfer, *Der Geburtenrückgang und seine Bekämpfung. Die Lebensfrage des deutschen Volkes* (Berlin: R. Schoetz, 1929); Friedrich Burgdörfer, *Volk ohne Jugend: Geburtenschwind und Überalterung des deutschen Volkskörpers; ein Problem der Volkswirtschaft, der Sozialpolitik, der nationalen Zukunft* (Berlin: K. Vowinckel, 1932). On European anxieties about population decline in general, see Michael S. Teitelbaum and J. M. Winter, *The Fear of Population Decline* (Orlando, FL: Academic Press, 1985).

18. On the rhetorical strategy of negation and "empty spaces" in colonial discourse, see David Spurr, *The Rhetoric of Empire: Colonial Discourse in Journalism, Travel Writing, and Imperial Administration* (Durham, NC: Duke University Press, 1993), 92–108. On the growing importance of global "wastelands" for allegedly overpopulated European countries in the 1920s and 1930s, see Alison Bashford, *Global Population: History, Geopolitics, and Life on Earth*, Columbia Studies in International and Global History (New York: Columbia University Press, 2014), 133–53; ADM NL 92/134: Map of North Africa, n.d.; Herman Sörgel, "Neugestaltung der Erdoberfläche durch den Ingenieur," in *Die Welt im Fortschritt: Gemeinverständliche Bücher des Wissens und Forschens der Gegenwart*, vol. 2 (Berlin: F. A. Herbig, 1935), 34–36; Sörgel, *Atlantropa*, 42.

19. See David M. Pletcher, *The Diplomacy of Trade and Investment: American Economic Expansion in the Hemisphere, 1865–1900* (Columbia: University of Missouri Press, 1998), 334–35; Mark Reisler, *By the Sweat of Their Brow: Mexican Immigrant Labor in the United States, 1900–1940* (Westport, CT: Greenwood, 1976), 205; "Forming Company to Irrigate Sahara," *New York Times*, April 27, 1929; "Vast Lake Planned to Irrigate Sahara," *New York Times*, September 16, 1928.

20. Sörgel, "Panropa-Projekt," 216–18; Raoul Heinrich Francé, *Das Land der Sehnsucht: Reisen eines Naturforschers im Süden* (Berlin: J.H.W. Dietz Nachfolger, 1925), 161–73.

21. Sörgel, *Die drei großen "A,"* 44–51. After reading a British survey on global soil erosion, Sörgel upped his estimate about the growth rate of the Sahara: for the survey, see G. V. Jacks and R. O. Whyte, *The Rape of the Earth: A World Survey of Soil Erosion* (London: Faber and Faber, 1939), 61–75; ADM NL 92/86: Sörgel, "Austrocknung in Afrika" (handwritten note), n.d. Diana K. Davis, *Resurrecting the Granary of Rome: Environmental History and French Colonial Expansion in North Africa* (Athens: Ohio University Press, 2007). It was only in the 1980s that the "granary of Rome" thesis began to be seriously questioned; see Brent D. Shaw, "Climate, Environment, and History: The Case of Roman North Africa," in *Climate and History: Studies in Past Climates and Their Impact on Man*, ed. T. M. L. Wigley, M. J. Ingram, and G. Farmer (Cambridge: Cambridge University Press, 1981), 379–403.

22. Ellsworth Huntington, *Civilization and Climate*, 3rd ed. (New Haven, CT: Yale University Press, 1924); C. E. P. Brooks, *The Evolution of Climate* (London: Benn Brothers, 1922); C. E. P. Brooks, *Climate Through the Ages: A Study of the Climatic Factors and Their Variations* (London: E. Benn, 1926); see also Richard Grove and Vinita Damodaran, "Imperialism, Intellectual Networks, and Environmental Change: Origins and Evolution of Global Environmental History, 1676–2000," *Economic and Political Weekly* 41, nos. 41–42 (October 2006): 4347–49. On the "epochal significance" of the Dust Bowl for renewed environmental anxieties in the 1930s, see Joachim Radkau, *The Age of Ecology: A Global History*, trans. Patrick Camiller (Cambridge: Polity, 2014), 46–61; see also Paul B. Sears, *Deserts on the March* (Norman: University of Oklahoma Press, 1935); Diana K. Davis, *The Arid Lands: History, Power, Knowledge*, History for a Sustainable Future (Cambridge, MA: MIT Press, 2015), 117–42.

23. Edward Percy Stebbing, *The Forests of India*, 3 vols. (London: Bodley Head, 1922). For a discussion of some parts of this book, see Greg Barton, *Empire Forestry and the Origins of Environmentalism*, Cambridge Studies in Historical Geography 34 (Cambridge: Cambridge University Press, 2002); Edward Percy Stebbing, "The Encroaching Sahara: The Threat to the West African Colonies," *Geographical Journal* 85, no. 6 (1935): 506–19. For the contemporary discussion of this lecture, see Percy Cox et al., "The Encroaching Sahara: The Threat to the West African Colonies: Discussion," *Geographical Journal* 85, no. 6 (June 1935): 519–24. Stebbing still propagated his views in the 1950s; see Edward Percy Stebbing, *The Creeping Desert in the Sudan and Elsewhere in Africa, 15° to 13° Latitude* (Khartoum, Sudan: McCorquodale, 1953). On the misreading of the African landscape by Stebbing, see Michael Mortimore, *Roots in the African Dust: Sustaining the Sub-Saharan Drylands* (Cambridge: Cambridge University Press, 1998), 19–21.

24. Edward Percy Stebbing, "The Threat of the Sahara," *Journal of the Royal African Society* 36, no. 145 (October 1937): 3–35.

25. E. William Bovill, "The Encroachment of the Sahara on the Sudan," *Journal of the Royal African Society* 20, no. 79 (April 1, 1921): 174–85; G. T. Renner, "A Famine Zone in Africa: The Sudan," *Geographical Review* 16, no. 4 (October 1, 1926): 583–96; William Malcolm Hailey, *An African Survey: A Study of Problems Arising in Africa South of the Sahara* (London: Oxford University Press, 1938); E. Barton Worthington, *The Ecological Century: A Personal Appraisal* (Oxford: Clarendon, 1983), 36–37; ADM NL 92/193: "Die Sahara ein Blumengarten?," *8 Uhr-Blatt*, July 2, 1935.

26. ADM NL 92/32: Sörgel to Paolo Vinassa de Regny, October 26, 1931.

27. ADM NL 92/140: "Die Geschichte Atlantropas," *Atlantropa Mitteilungen* 14 (1948) and Siegwart to the *Münchener Neueste Nachrichten*, June 8, 1929.

28. While it is difficult to predict the geological effects Atlantropa would have had, a large redistribution of water would lead to a different surface load of the earth and thus could potentially alter the rotation axis of the Earth; see Kurt Lambeck, *The Earth's Variable Rotation: Geophysical Causes and Consequences* (Cambridge: Cambridge University Press, 1980), 50–52, 163–66; ADM NL 92/232: Siegwart to Sörgel, March 14, 1930.

29. ADM NL 92/86: Sörgel to the Kölnische Volkszeitung, January 1, 1931; Sörgel, *Atlantropa*, 65–70; on reservoir-induced seismicity; see Patrick McCully, *Silenced Rivers: The Ecology and Politics of Large Dams* (London: Zed Books, 2001), 112–15; C. H. Scholz, *The Mechanics of Earthquakes and Faulting*, 2nd ed. (Cambridge: Cambridge University Press, 2002), 344–48; Michael Manga and Chi-yuen Wang, *Earthquakes and Water* (Heidelberg: Springer, 2010), 128–30.

30. ADM NL 92/231: Siegwart to Sörgel, October 14, 1932, Siegwart to Sörgel, November 10, 1933; Siegwart to Sörgel, July 15, 1934.

31. Sörgel and Siegwart, "Erschliessung Afrikas"; ADM NL 92/231: Siegwart to Sörgel, November 10, 1933.

32. ADM NL 92/132: Sörgel, "Kongo und Afrikawerke" (handwritten note), n.d.; ADM NL 92/86: Sörgel, "Die Bewässerung der Kalahari-Wüste" (handwritten note), n.d.; ADM NL 92/196: "Ein Kraftwerk in der Wüste: Unterredung mit dem Wüstenforscher Graf L. E. von Almasy," *Deutsche Bergwerkszeitung* 275, November 23, 1940.

33. F. Gessert, "Wüsten und Pole," *Die Umschau* 46 (1925): 907; August Wendler, *Das Problem der technischen Wetterbeeinflussung*, Probleme der kosmischen Physik 9 (Hamburg: Henri Grand, 1927), 8.

34. Paul Sokolowski, *Die Versandung Europas: . . . eine andere, große russische Gefahr* (Berlin: Deutsche Rundschau, 1929); Sörgel, *Atlantropa*, 84.

35. Sörgel, *Mittelmeer-Senkung, Sahara-Bewässerung*, 32.

36. Sörgel, *Atlantropa*, 66–67. It is not entirely clear what drove Sörgel to make his claims about a cooling Europe. From 1918 to 1933, only the winter of 1929 had been a particularly cold one in Central Europe. There was also no discernible general trend toward colder average temperatures in Germany or Central Europe during that period; see Frank Sirocko, Heiko Brunck, and Stephan Pfahl, "Solar Influence on Winter Severity in Central Europe," *Geophysical Research Letters* 39, no. 16 (2012); Jörg Rapp and Christian-Dietrich Schönwiese, *Climate Trend Atlas of Europe Based on Observations, 1891–1990* (Dordrecht: Kluwer, 1997), 30–32; ADM NL 92/193: Sörgel, "Klimaverbesserung und Atlantropa," *Rathenower Zeitung*, June 28, 1933, 193. The idea of large oceanic dams to regulate currents and change climatic conditions had already been discussed by the American Carroll Riker, who proposed a dam jutting out from Newfoundland to cut off the cold Labrador Current and divert the Gulf Stream; see Carroll Livingston Riker, *Power and Control of the Gulf Stream: How It Regulates the Climates, Heat and Light of the World* (Brooklyn: Baker & Taylor, 1912); Carroll Livingston Riker, *Conspectus of Power and Control of the Gulf Stream* (Brooklyn: C. L. Riker, 1913). The potential shift of oceanic currents due to large engineering systems had also evoked anxieties over detrimental climatic changes. This was one of the criticisms of MacKenzie's Sahara Sea project; see Edwin E. Slosson, "Plans to Restore the City of Brass," *Science News-Letter* 14, no. 401 (December 15, 1928): 313.

37. ADM NL 92/231: Siegwart to Sörgel, July 15, 1934; Sörgel, *Die drei großen "A,"* 23–26.

38. Cf. Gall, *Das Atlantropa-Projekt*, 14, 54; ADM NL 92/191: "Ein gigantischer Plan," *Neuköllner Tageblatt*, April 11, 1929.

Chapter 5

1. Herman Sörgel, *Atlantropa* (Zurich: Fretz & Wasmuth, 1932), 6, emphasis added.

2. ADM NL 92/92: Sörgel, "Exposé zur Deutschen Atlantropa Weltausstellung Berlin 1937," n.d.; ADM NL 92/194: Sörgel, "Atlantropa-Weltausstellung," *Die Sonntags-Zeitung Stuttgart*, August 9, 1936; ADM NL 92/195: Sörgel, "Der Friede bricht aus! Totaler Krieg oder totaler Friede?," *Deutsche Freiheit*, December 1937; Herman Sörgel, *Die drei großen "A": Großdeutschland und italienisches Imperium, die Pfeiler Atlantropas* (Munich: Piloty & Loehle, 1938), 119–25.

3. Alexander Gall, *Das Atlantropa-Projekt: Die Geschichte einer gescheiterten Vision. Herman Sörgel und die Absenkung des Mittelmeers* (Frankfurt am Main: Campus, 1998), 38–39; ADM NL 92/132 and 135; ADM NL 92/49: Sörgel, "Anregung zu einer Atlantropa-Sinfonie (Friedens-Sinfonie)" (handwritten note), n.d.

4. For more information on Behrens, see Stanford Anderson, *Peter Behrens and a New Architecture for the Twentieth Century* (Cambridge, MA: MIT Press, 2000); ADM NL 92/140: "Die Geschichte Atlantropas," *Atlantropa Mitteilungen* 15/16 (1948); Gall, *Das Atlantropa-Projekt*, 34–36; Wolfgang Voigt, "Ein hypertrophes Projekt der Moderne. Herman Sörgel und sein Kontinent Atlantropa," *Bauwelt* 80, no. 18/19 (1991): 938–64.

5. Based on Sörgel's Atlantropa archive, Gall counts 466 articles on Atlantropa from 1929 to 1933 in the German-speaking news media; see Gall, *Das Atlantropa-Projekt*, 38. Even in the United States, Atlantropa was sporadically covered by the press; see, e.g., "Proposes to Lower Mediterranean Sea," *New York Times*, April 14, 1929; "Huge Dikes Are Now Proposed to Dry up the Mediterranean," *New York Times*, December 29, 1929; "Atlantropa Plan Would Dam Straits," *Washington Post*, May 26, 1951; Richard Hennig, "Sörgels Plan eines Binnenschiffahrtsweges durch die Sahara," *Technik und Wirtschaft* 29 (1936): 180–82. For Hennig's contribution to the Atlantis debate, see Richard Hennig, "Zur neuen Borchardt-Hermannschen Atlantis- und Tartessoshypothese," *Petermanns Mitteilungen aus Justus Perthes' Geographischer Anstalt* 73 (1927): 282–84.

6. On the role of engineers—and in particular hydro-engineers—as "heroes of the modern age" in the early twentieth century, see Denis E. Cosgrove, "An Elemental Division: Water Control and Engineered Landscape," in *Water, Engineering, and Landscape: Water Control and Landscape Transformation in the Modern Period*, ed. Denis E. Cosgrove and Geoffrey E. Petts (London: Belhaven Press, 1990), 7–8; Hennig, "Sörgels Plan eines Binnenschiffahrtsweges durch die Sahara," 182.

7. H. Bode, Robert Potonié, and Karl Haushofer, "Was sagt die Wissenschaft zum Projekt der Mittelmeersenkung?," *Reclams Universum* 46, no. 12 (1929): 258. For a study of Haushofer beyond foregone conclusions about his role in shaping and propagating National Socialist ideology, see Frank Ebeling, *Geopolitik: Karl Haushofer und seine Raumwissenschaft 1919–1945* (Berlin: Akademie Verlag, 1994). For a discussion of Haushofer's role in the Third Reich, see Holger H. Herwig, "Geopolitik: Haushofer, Hitler and Lebensraum," *Journal of Strategic Studies* 22, nos. 2–3 (1999): 218–41.

8. ADM NL 92/198: Karl Haushofer, "Panropa," *Hamburger Fremdenblatt*, April 25, 1931; ADM NL 92/100: Handwritten note by Sörgel, n.d. (ca. 1940).

9. See David Thomas Murphy, *The Heroic Earth: Geopolitical Thought in Weimar Germany, 1918–1933* (Kent, OH: Kent State University Press, 1997); Heike Wolter, *"Volk ohne Raum"— Lebensraumvorstellungen im geopolitischen, literarischen und politischen Diskurs der Weimarer Republik: eine Untersuchung auf der Basis von Fallstudien zu Leben und Werk Karl Haushofers, Hans Grimms und Adolf Hitlers* (Münster: Lit, 2003); Klaus Kost, *Die Einflüsse der Geopolitik auf Forschung und Theorie der politischen Geographie von ihren Anfängen bis 1945*, Bonner Geographische Abhandlungen 76 (Bonn: Ferd. Dümmlers Verlag, 1988), 112–20; Sörgel, *Atlantropa*, 82; James Fairgrieve, *Geography and World Power* (New York: Dutton, 1917), 339–40; cf. Johannes Walther, *Das Gesetz der Wüstenbildung in Gegenwart und Vorzeit*, 2nd ed. (Leipzig: Quelle & Meyer, 1912), 99.

10. On the history of the term *Lebensraum*, see Woodruff D. Smith, "Friedrich Ratzel and the Origins of Lebensraum," *German Studies Review* 3, no. 1 (February 1, 1980): 51–68; Woodruff D. Smith, *Politics and the Sciences of Culture in Germany, 1840–1920* (New York: Oxford University Press, 1991), 219–33.

11. See Friedrich Ratzel, *Der Lebensraum: Eine biogeographische Studie* (Tübingen: H. Laupp, 1901), 1.

12. Ratzel's basic understanding of the workings of history is, for instance, similar to John McNeill's definition of "material environmental history," See John Robert McNeill, "The State of the Field of Environmental History," *Annual Review of Environment and Resources* 35, no. 1 (2010): 347.

13. Friedrich Ratzel, *Anthropo-Geographie: Grundzüge der Anwendung der Erdkunde auf die Geschichte*, vol. 1 (Stuttgart: J. Engelhorn, 1882), 32.

14. Ratzel, *Anthropo-Geographie*, 1:467; Friedrich Ratzel, *Anthropogeographie: Die geographische Verbreitung des Menschen*, vol. 2 (Stuttgart: J. Engelhorn, 1891), xxxiv–xxxv; see also Gerhard H. Müller, "Das Konzept der 'Allgemeinen Biogeographie' von Friedrich Ratzel (1844–1904): eine Übersicht," *Geographische Zeitschrift* 74, no. 1 (March 1, 1986): 3–14; Johannes Steinmetzler, *Die Anthropogeographie Friedrich Ratzels und ihre ideengeschichtlichen Wurzeln*, Bonner geographische Abhandlungen 19 (Bonn: Selbstverlag des Geographischen Instituts der Universität Bonn, 1956), 43–45.

15. Herman Sörgel, "Das Panropa-Projekt als städtebauliche Darstellungsstudie," *Baumeister* 29, no. 5 (1931): 216.

16. Sörgel, *Atlantropa*, 137; cf. Herman Sörgel, "Neugestaltung der Erdoberfläche durch den Ingenieur," in *Die Welt im Fortschritt: Gemeinverständliche Bücher des Wissens und Forschens der Gegenwart*, vol. 2 (Berlin: F. A. Herbig, 1935), 65.

17. Sörgel, *Atlantropa*, 65; ADM NL 92/132: Drawing by Sörgel, n.d.; Sörgel, *Die drei großen "A,"* 115; Sörgel, "Neugestaltung der Erdoberfläche," 21.

18. Sörgel, "Neugestaltung der Erdoberfläche," 21–24; Sörgel, *Atlantropa*, 4–8, 39–48; Sörgel believed that a lake had existed in place of the Mediterranean Sea connected to the Atlantic Ocean up to about fifty thousand years ago. Current research suggests that a flooding event of the Mediterranean basin did indeed take place, but happened more than five million years ago in the Late Miocene; see D. Garcia-Castellanos et al., "Catastrophic Flood of the Mediterranean after the Messinian Salinity Crisis," *Nature* 462, no. 7274 (2009): 778–81.

19. Sörgel, like Roudaire before him, probably attempted to make his project appear less radical with his rhetoric of technology as an extension of nature. This view, however, was also more widespread. As Richard White has argued, technology could seem opposed to nature from one vantage point, and as a manifestation of natural forces from another; see Richard White, *The Organic Machine* (New York: Hill & Wang, 1995), 30; Sörgel, *Atlantropa*, 34, 70; Sörgel, "Neugestaltung der Erdoberfläche," 20.

20. ADM NL 92/231: Handwritten note by Sörgel, July 17, 1930; Sörgel, *Atlantropa*, 33, 68.

21. Sörgel, *Die drei großen "A,"* 41; Sörgel, "Panropa-Projekt," 218. On the causes and effects of the economic crisis in Germany from 1930 to 1934, see Harold James, *The German Slump: Politics and Economics, 1924–1936* (Oxford: Clarendon, 1986).

22. Karl-Heinz Ludwig, *Technik und Ingenieure im Dritten Reich* (Düsseldorf: Droste, 1974), 49–50. For an overview of the development of engineering as a profession in Germany, see Kees Gispen, *New Profession, Old Order: Engineers and German Society, 1815–1914* (Cambridge: Cambridge University Press, 1989); Sörgel, "Neugestaltung der Erdoberfläche," 17, emphasis original. On ideas of the use of technology to unify political entities, or "techno-globalism," see David Edgerton, *The Shock of the Old: Technology and Global History since 1900* (Oxford: Oxford University Press, 2007), 113–17.

23. Wayne W. Parrish, *Technokratie—die neue Heilslehre* (Munich: R. Piper, 1933). For the original English version, see Wayne W. Parrish, *An Outline of Technocracy* (New York: Farrar & Rinehart, 1933). On the idea of economic energetics, see Howard Scott, *Technocracy, a Thermodynamic Interpretation of Social Phenomena* (New York: Technocracy Inc., 1932); Stefan Willeke, *Die Technokratiebewegung in Nordamerika und Deutschland zwischen den Weltkriegen* (Frankfurt am Main: P. Lang, 1995), 131. On the history of the technocratic movement in the United States, see William E. Akin, *Technocracy and the American Dream: The Technocrat Movement, 1900–1941* (Berkeley: University of California Press, 1977); Gall, *Das Atlantropa-Projekt*, 97.

24. Oswald Spengler, *Der Untergang des Abendlandes: Umrisse einer Morphologie der Weltgeschichte* (Munich: Beck, 1931). On Spengler's theories of cultural decline and an "end of history," see Samir Osmančević, *Oswald Spengler und das Ende der Geschichte* (Vienna: Turia Kant, 2007), 120–22; Wolfgang Krebs, *Die imperiale Endzeit: Oswald Spengler und die Zukunft der abendländischen Zivilisation* (Berlin: Rhombos, 2008), 47–78; Sörgel, *Die drei großen "A,"* 91; Sörgel, *Atlantropa*, 104.

25. Sörgel's reading was, in fact, quite close to Jeffrey Herf's interpretation of Spengler as a technology enthusiast; see Jeffrey Herf, *Reactionary Modernism: Technology, Culture, and Politics in Weimar and the Third Reich* (Cambridge: Cambridge University Press, 1984), 49–69; Oswald Spengler, *Der Mensch und die Technik* (Paderborn: Voltmedia, 2007), 39, 75.

26. Spengler, *Der Mensch und die Technik*, 82. Spengler's attempt to court the political right is maybe most conspicuous in Oswald Spengler, *Preussentum und Sozialismus* (Munich: C. H. Beck, 1921); Sörgel, *Atlantropa*, 70. On the importance of the concept of energy in European thought in the nineteenth and early twentieth centuries, see Anson Rabinbach, *The Human Motor: Energy, Fatigue, and the Origins of Modernity* (Los Angeles: University of California Press, 1992). In Thomas Rohkrämer's cataloguing of types of civilizational critique, Sörgel most closely corresponds to the third group, described as trying to solve any technological problems by the fine-tuning and full expression of technology itself: Thomas Rohkrämer, *Eine andere Moderne?: Zivilisationskritik, Natur und Technik in Deutschland 1880–1933* (Paderborn: Schöningh, 1999), 32–34.

27. ADM NL 92/34: Sörgel, "Ursymbole der Kulturkreise (Im Sinne Spenglers)" (manuscript), n.d.; Parrish, *Technokratie—die neue Heilslehre*, 13.

28. See Charles Maier, "Zwischen Taylorismus und Technokratie: Gesellschaftspolitik im Zeichen industrieller Rationalität in den zwanziger Jahren in Europa," in *Die Weimarer Republik: Belagerte Civitas*, ed. Michael Stürmer (Königstein: Verlagsgruppe Athenäum, Hain, Scriptor, Hanstein, 1980), 28. The debate between technological enthusiasts and doomsayers in Weimar was also not necessarily a debate over modernity as such. As Thomas Rohkrämer has pointed out, what is often perceived as "anti-modern" in interwar Germany was, in fact, often a search for a "different and better modernity" that would keep the positive effects, like advances in engineering, while doing away with the negative ones; see Rohkrämer, *Eine andere Moderne?*, 32; Detlev Peukert, *The Weimar Republic: The Crisis of Classical Modernity* (London: Penguin, 1991), 178–90.

29. Herman Sörgel and Bruno Siegwart, "Erschliessung Afrikas durch Binnenmeere: Saharabewässerung durch Mittelmeersenkung," *Beilage zum Baumeister* 3 (1935): 39; Joachim H. Schultze, "Die Tropen als Arbeitsfeld des Ingenieurs," *Zeitschrift des Vereins deutscher Ingenieure* 84, no. 49 (1940): 948; Karl Mannheim, *Man and Society in an Age of Reconstruction: Studies in Modern Social Structure*, ed. Edward Shils (London: K. Paul, Trench, Trubner & Co., 1940), 155.

30. ADM NL 92/446: Italian Consulate to Sörgel, September 27, 1929.

31. Cf. Alfred Rosenberg, *Der Mythus des 20. Jahrhunderts, eine Wertung der seelischgeistigen Gestaltenkämpfe unserer Zeit* (Munich: Hoheneichen-Verlag, 1935); ADM NL 92/20: Handwritten note by Sörgel, n.d. (ca. 1935); the text appeared as an article in Die Sonntags-Zeitung Stuttgart, August 9, 1936 (see ADM NL 92/194); ADM NL 92/193: Sörgel, "Aussprache," *Schule der Freiheit* 2, no. 16 (1934).

32. ADM NL 92/49: Sörgel, "Zur Kritik der Zeit" (handwritten note), n.d.; ADM NL 92/189: Sörgel, "Das Mittelmeer—Größte Kraftquelle der Zukunft," Wissen und Fortschritt, December 1929 Herman Sörgel, *Mittelmeer-Senkung, Sahara-Bewässerung (Panropa-Projekt)* (Leipzig: J. M. Gebhardt, 1929), 42; ADM NL 92/231: Sörgel, "Nachtrag zum Congo-Projekt" (manuscript), n.d.

33. ADM NL 92/231: Siegwart to Sörgel, July 15, 1934; Sörgel, *Atlantropa*, 103; Sörgel, *Die drei großen "A,"* 18–19, 60.

34. Sörgel, *Die drei großen "A,"* v, 16, 30.

35. Sörgel, *Atlantropa*, 79–84; Sörgel, *Die drei großen "A,"* vi, 17; ADM NL 92/195: Sörgel, "Vom Benzinauto zum Elektroauto," *Deutsche Freiheit* 6, no. 22 (1937): 4.

36. For an overview of Nazi efforts to create an autarkic economic system, see Eckart Teichert, *Autarkie und Großraumwirtschaft in Deutschland 1930–1939: Außenwirtschaftspolitische Konzeptionen zwischen Wirtschaftskrise und Zweitem Weltkrieg* (Munich: Oldenbourg, 1984); Tiago Saraiva and M. Norton Wise, "Autarky/Autarchy: Genetics, Food Production, and the Building of Fascism," *Historical Studies in the Natural Sciences* 40, no. 4 (2010): 419–28. Kost, *Die Einflüsse der Geopolitik*, 240ff.; Dietmar Petzina, *Autarkiepolitik im Dritten Reich* (Stuttgart: Deutsche Verlagsanstalt, 1968); Adam Tooze, *The Wages of Destruction: The Making and Breaking of the Nazi Economy* (New York: Penguin, 2008), 86–96.

37. Herbert von Obwurzer, *Selbstversorgung (Autarkie) im Dritten Reich* (Berlin: Nationaler Freiheitsverlag, 1933), 55; Helmut Maier, "'Weiße Kohle' versus Schwarze Kohle. Naturschutz und Ressourcenschonung als Deckmantel nationalsozialistischer Energiepolitik," *WerkstattGeschichte* 3 (1992): 33–38. On the Unternehmen Wüste, see BArch NS 3/823; Michael Grandt,

Unternehmen "Wüste"—Hitlers letzte Hoffnung: Das NS-Ölschieferprogramm auf der Schwäbischen Alb (Tübingen: Silberburg-Verlag, 2002); Christine Glauning, *Entgrenzung und KZ-System: Das Unternehmen "Wüste" und das Konzentrationslager in Bisingen 1944/45*, Geschichte der Konzentrationslager 1933–1945 7 (Berlin: Metropol, 2006).

38. Gall, *Das Atlantropa-Projekt*, 81–86; ADM NL 92/49: Sörgel to Köster (Department Head of the Reichsstelle für Raumordnung), n.d.

39. ADM NL 92/86: Sörgel, "Abgrenzungsmöglichkeit und Schutz gegen Asien" (handwritten note), n.d.; ADM NL 92/34: Sörgel, "Ost oder Süd" (manuscript), fall 1941; see also Peter Christensen, "Dam Nation: Imaging and Imagining the 'Middle East' in Herman Sörgel's Atlantropa," *International Journal of Islamic Architecture* 1, no. 2 (August 17, 2012): 325–46; Sörgel, *Die drei großen "A,"* 29–30, 34–37.

40. ADM NL 92/86: Sörgel, "Nord-Süd" (handwritten note), n.d.; ADM NL 92/195: Sörgel, 'Weiß oder Farbig?," *Deutsche Freiheit* 29 (October 1937); Michael Keevak, *Becoming Yellow: A Short History of Racial Thinking* (Princeton, NJ: Princeton University Press, 2011); Sörgel, "Panropa-Projekt," 219. For German perceptions of China and its population, see Mechthild Leutner, "Deutsche Vorstellungen über China und Chinesen und über die Rolle der Deutschen in China 1890–1945," in *Von der Kolonialpolitik zur Kooperation: Studien zur Geschichte der deutsch-chinesischen Beziehungen*, ed. Heng-yü Kuo, Berliner China-Studien 13 (Munich: Minerva, 1986), 401–43; ADM NL 92/34: Sörgel, "Ursymbole der Kulturkreise"; cf. ADM NL 92/110: Sörgel, "Fragenbeantwortung" (typescript), February 12, 1926.

41. See ADM NL 92/100; ADM NL 92/231: Siegwart to Sörgel, November 25, 1941; ADM NL 92/140: "Die Geschichte Atlantropas," *Atlantropa Mitteilungen* 22 (1949); ADM NL 92/34: Sörgel, "Vor dem Richterstuhl der Zeit," n.d.

42. John Knittel had been close to the Nazi leadership in the 1930s and was subsequently shunned by many of his colleagues after the fall of the Third Reich. Already in 1939, Knittel had written an homage to Atlantropa; see John Knittel, *Amadeus* (Berlin: W. Krüger, 1939). Knittel also wrote the foreword to Sörgel's 1948 publication on Atlantropa; see Herman Sörgel, *Atlantropa. Wesenszüge eines Projekts* (Stuttgart: Behrendt, 1948). In 1947, Knittel had attempted to establish connections to the newly founded UNESCO, but his attempt to contact Director-General Julian Huxely had been unsuccessful; see ADM NL 92/140: *Atlantropa Mitteilungen* 6 (1947). The UN magazine ran a short but very positive article on Sörgel's project in 1948: "A Dam at Gibraltar," *UN World* 2 (May 1948): 48–49. Karl Theens became the "head of the research division" of the Atlantropa Institute and published his own monograph on the project in 1949; see Karl Theens, *Der Sörgel-Plan: Afrika + Europa = Atlantropa* (Bielefeld: Küster, 1949). Later, Theens became the curator of the Faust Museum in Knittlingen; see Karl Hochwald, "Zehn Jahre Faust-Gedenkstätte in Knittlingen," in *Faust im zwanzigsten Jahrhundert: Festschrift für Karl Theens zum sechzigsten Geburtstag*, ed. Henri Clemens Birven, Dietmar Theens, and Karl Weisert (Knittlingen: Stadtverwaltung, 1964), 47–49.

43. ADM NL 92/31: Sörgel, "Es schwinden, es fallen die leidenden Menschen blindlings von einer Stunde zur anderen" (manuscript), n.d. (ca. 1946–49); ADM NL 92/52: Sörgel, "Idee eines Zwölferrates zur Vorbereitung von Atlantropa" (handwritten note), n.d. (ca. 1946–49).

44. Anton Zischka, *Brot für zwei Milliarden Menschen* (Leipzig: W. Goldmann, 1938), 7, 298, 304–5; Anton Zischka, *Afrika, Europas Gemeinschaftsaufgabe Nr. 1* (Oldenburg: Gerhard Stalling, 1951), 13–14, 60–70, 86–94, 109.

45. See Stephen Brain, "The Great Stalin Plan for the Transformation of Nature," *Environmental History* 15, no. 4 (October 1, 2010): 670–700; Klaus Gestwa, *Die Stalinschen Großbauten des Kommunismus. Sowjetische Technik- und Umweltgeschichte, 1948–1967* (Munich: Oldenbourg, 2010), 130ff.; Maya K. Peterson, *Pipe Dreams: Water and Empire in Central Asia's Aral Sea Basin* (Cambridge: Cambridge University Press, 2019); Marc Elie, "Desiccated Steppes: Droughts and Climate Change in the USSR, 1960s–1980s," in *Eurasian Environments*, ed. Nicholas B. Breyfogle (Pittsburgh: University of Pittsburgh Press, 2018), 75–94. On the connection between climate change ideas and climate engineering in the Soviet Union, see Jonathan D. Oldfield, "Climate Modification and Climate Change Debates Among Soviet Physical Geographers, 1940s–1960s," *Wiley Interdisciplinary Reviews: Climate Change* 4, no. 6 (November 1, 2013): 513–24; "Projekt Dawidow: Ums Leben kommen," *Der Spiegel*, May 11, 1950; ADM NL 92/6: Sörgel, "Rußland 'baut' ein Meer," *Deutsche Presse Korrespondenz* 24 (1950): 2–3; BArch B 136/6063: "Angebote aus der UdSSR für Soergel," *Die Neue Zeitung: Die Amerikanische Zeitung in Deutschland*, April 12, 1951.

46. Sörgel sensed the trend toward nuclear energy early. As early as 1936, Sörgel had denounced the attempts to gain energy from "atomic shattering" as far-fetched and described hydropower as the only solution to the imminent energy crisis: ADM NL 92/194: Sörgel, "Die Welt ohne Kohle," Sonntags-Zeitung, October 4, 1936; twelve years later, his polemic had become sharper, highlighting the destructive power of nuclear technology and the difficulties of garnering usable energy from atomic processes: ADM NL 92/73: Sörgel, "Die Welt ohne Kohle: Atlantropa und die künftige Energiewirtschaft," *Natur und Technik* 19 (1948): 283–90; Herman Sörgel, *Idee und Macht: Ein Sang von Atlantropa. Zum 25-jährigen Bestehen Atlantropas* (Oberstdorf, 1950); ADM NL 92/140: "Wichtige Mitteilung," *Atlantropa Mitteilungen* 15/16 (1949); ADM NL 92/6; Kurt Hiehle, "Ist Atlantropa nur eine Utopie?," *Blick in die Wissenschaft* 1 (1948): 236–38; ADM NL 92/6: Kurt Hiehle, "Atlantropa ist keine Utopie aber der Gibraltardamm ist ein Irrweg," *Frankfurter Allgemeine Zeitung*, July 15, 1950.

47. BArch B 136/6063: Loeffelholz to the Atlantropa Institute, Munich, July 20, 1951; see also Gall, *Das Atlantropa-Projekt*, 45–47.

48. Hans Aburi, "Das Ende einer großen Idee," *Neue Politik* 5, no. 33 (1960): 8–9; J. Fritzsche, *Atlantropa, die Stimme der eurafrikanischen Friedensbürger*, 1 (Starnberg: J. Fritzsche Buch- und Zeitschriftenvertrieb, 1963).

49. ADM NL 92/31: Sörgel, "Es schwinden, es fallen die leidenden Menschen."

50. ADM NL 92/196: W. Harnisch, "Das Ende der Kohle," *Deutsche Bergwerks-Zeitung*, March 25, 1941; ADM NL 92/86: Sörgel, "Austrocknung in Afrika" and "Literatur über Afrika" (handwritten notes), n.d.; A. E. Johann, *Groß ist Afrika: Vom Kap über den Kongo zur Westküste* (Berlin: Deutscher Verlag, 1939), 153–56; ADM NL 92/196: "Kann der Mensch das Klima lenken?," *Westdeutscher Beobachter*, May 13, 1942.

51. ADM NL 92/140: "Atlantropa verändert die Geographie Europa-Afrikas," *Atlantropa Mitteilungen* 25 (1949); Anton Metternich, *Die Wüste droht: Die gefährdete Nahrungsgrundlage der menschlichen Gesellschaft* (Bremen: F. Trüjen, 1947); on Metternich, see Franz Dreyhaupt, *Frühe Umwelt-Warner—Rufer in der Wüste?: Ein Beitrag zur Umweltgeschichte* (Düren: F. J. Dreyhaupt, 2008), 12–15; ADM NL 92/52: Sörgel, "Der Atlantropa-Plan mit besonderer Berücksichtigung von Wasser und Boden" (manuscript), n.d. Theens developed his own ideas of installing huge air conditioning systems in Africa to make the continent habitable for Europeans: Theens, *Der Sörgel-Plan*, 42.

52. Gall, *Das Atlantropa-Projekt*, 130; ADM NL 92/140: "Atlantropa verändert die Geographie Europa-Afrikas," *Atlantropa Mitteilungen* 25 (1949); C. Troll, J. van Eimern, and W. Daume, "Herman Sörgels 'Atlantropa' in geographischer Sicht," *Erdkunde* 4 (1950): 177–88.

53. See Hermann Flohn, "Die Tätigkeit des Menschen als Klimafaktor," *Zeitschrift für Erdkunde* 9 (1941): 13–22; for the English version of Flohn's standard work on climate, see Hermann Flohn, *Climate and Weather*, trans. B. V. de G. Walden (London: Weidenfeld & Nicolson, 1969). On Flohn, see Matthias Heymann and Dania Achermann, "From Climatology to Climate Science in the Twentieth Century," in *The Palgrave Handbook of Climate History*, ed. Sam White, Christian Pfister, and Franz Mauelshagen (London: Palgrave Macmillan, 2018), 605–32; Fritz Jaeger, Hermann Flohn, and A. Schmauß, "Bemerkungen zum Atlantropa-Projekt," *Erdkunde* 5 (1951): 179–80.

54. David Blackbourn, *The Conquest of Nature: Water, Landscape, and the Making of Modern Germany* (New York: Norton, 2006), 278–93; Alwin Seifert, "Die Versteppung Deutschlands," in *Im Zeitalter des Lebendigen. Natur, Heimat, Technik* (Planegg: Müller, 1943), 24–51; Sörgel, *Die drei großen "A*," 23–26; ADM NL 92/195: Sörgel, "Blut und Boden," *Deutsche Freiheit*, August 1937.

55. Kurt Hiehle, *Vom kommenden Zeitalter der künstlichen Klimagestaltung* (Heidelberg: Brausdruck, 1947), 15.

Chapter 6

1. Frederick the Great's quote is cited in Hans Walter Flemming, *Wüsten, Deiche und Turbinen: Das große Buch von Wasser und Völkerschicksal* (Göttingen: Musterschmidt, 1957), 187; Heinrich Wiepking-Jürgensmann, "Friedrich der Große und Wir," *Die Gartenkunst* 33, no. 5 (1920): 70–71. Wiepking used his wife's last name Jürgensmann only until 1945. For the sake of convenience, I refer to Wiepking by his own last name only.

2. On the contemporary image of Frederick the Great, see Eva Giloi, *Monarchy, Myth, and Material Culture in Germany 1750–1950* (Cambridge: Cambridge University Press, 2011), 361–62. On the changing image of Frederick the Great in the two postwar Germanies, see Hans Dollinger, *Friedrich II. von Preussen: Sein Bild im Wandel von zwei Jahrhunderten* (Munich: List, 1986), 193–216; Frank-Lothar Kroll, "Friedrich der Große," in *Deutsche Erinnerungsorte*, ed. Etienne François and Hagen Schulze, 3 vols. (Munich: Beck, 2001), 3:620–35. The image of Frederick the Great as a man of action rather than a philosopher was common in Nazi ideology; see Ernst Adolf Dreyer and Heinz W. Siska, eds., *Kämpfer, Künder, Tatzeugen, Gestalter deutscher Grösse*, vol. 1 (Munich: Zinnen, 1942), 182–84, 203–5; Wiepking-Jürgensmann, "Friedrich der Große und Wir," 77.

3. For a comprehensive biography of Wiepking, see Ursula Kellner, "Heinrich Friedrich Wiepking (1871–1973): Leben, Lehre und Werk" (PhD diss., Universität Hannover, 1998).

4. Heinrich Wiepking-Jürgensmann, "Gegen den Steppengeist," *Das Schwarze Korps*, October 15, 1942, 4; NLSO Dep 72b/116: Wiepking, "Der neue Garten," speech, June 5, 1923, n.p.

5. There is now a large body of literature on Nazi plans for a complete reorganization of the occupied areas in the east; for a general overview of the genocidal resettlement plans, see Götz Aly, *"Endlösung": Völkerverschiebung und der Mord an den europäischen Juden* (Frankfurt am Main: S. Fischer, 1995); for the English version, see Götz Aly, *"Final Solution": Nazi Population Policy and the Murder of the European Jews*, trans. Belinda Cooper and Allison Brown (London: Arnold, 1999); Mark Mazower, *Hitler's Empire: How the Nazis Ruled Europe* (New York:

Penguin, 2008), 211–22. Martin Broszat, *Nationalsozialistische Polenpolitik 1939–1945*, Schriften-reihe der Vierteljahreshefte für Zeitgeschichte 2 (Stuttgart: Deutsche-Verlags-Anstalt, 1961); Robert Lewis Koehl, *RKFDV: German Resettlement and Population Policy, 1939–1945: History of the Reich Commission for the Strengthening of Germandom* (Cambridge, MA: Harvard University Press, 1957); Timothy Snyder, *Black Earth: The Holocaust as History and Warning* (New York: Tim Duggan Books, 2015). See also Timothy Snyder, "The Next Genocide," *New York Times*, September 12, 2015, sec. Opinion. On Nazi projects of environmental transformation, see David Blackbourn, *The Conquest of Nature: Water, Landscape, and the Making of Modern Germany* (New York: Norton, 2006), 253–310. On the environmental dimension of the Nazi "blood and soil" ideology, see Mark Bassin, "Blood or Soil? The Völkisch Movement, the Nazis, and the Legacy of Geopolitik," in *How Green Were the Nazis? Nature, Environment, and Nation in the Third Reich*, ed. Franz-Josef Brüggemeier, Mark Cioc, and Thomas Zeller, 1st ed., Ohio University Press Series in Ecology and History (Athens: Ohio University Press, 2005), 204–42.

6. Mineau's study of SS ideology, for example, refers to the influence of the biological sciences, but does not touch on influences from other fields of science; see André Mineau, *SS Thinking and the Holocaust* (Amsterdam: Rodopi, 2012). Similarly, Enzo Traverso's essay on Nazi ideology contains a short chapter on "Lebensraum," but classifies eugenics and racism as the "motor" of the "expansionist policy"; see Enzo Traverso, *The Origins of Nazi Violence*, trans. Janet Lloyd (New York: New Press, 2003), 74. George Mosse's classical study on the intellectual origins of National Socialism devotes a whole chapter to racism, but has little to say on the influence of geology or even of geography: George L. Mosse, *The Crisis of German Ideology: Intellectual Origins of the Third Reich* (New York: Howard Fertig, 1998), 88–107. Rather than using a derivation from the term for desert (*Wüste*), German authors in the 1930s and 1940s preferred the term *Versteppung*. One reason behind this choice is that *Verwüstung* already denoted (and still denotes) destruction through warfare and the abandonment of settlements, similar to the English term "devastation"; see, for example, Rudolf Bergmann, "Quellen, Arbeitsverfahren und Fragestellungen der Wüstungsforschung," *Siedlungsforschung: Archäologie, Geschichte, Geographie* 12 (1994): 35–68. *Versteppung* as used by German authors in the 1920s to 1940s described the erosion of soils, the change of the climate toward dryer average conditions, and the disappearance of plant cover. This is very close to what would be described by the term "desertification" today, but referred almost exclusively to landscapes in the east. On continuities from the Wilhelmine colonial era to Nazi plans of imperial conquest, see Shelley Baranowski, *Nazi Empire: German Colonialism and Imperialism from Bismarck to Hitler* (Cambridge: Cambridge University Press, 2011).

7. See Jürgen Zimmerer, "The Birth of the Ostland out of the Spirit of Colonialism: A Postcolonial Perspective on the Nazi Policy of Conquest and Extermination," *Patterns of Prejudice* 39, no. 2 (2005): 197–219. For Nazi Germany as an imperial power with far-reaching colonial aims, see also Mazower, *Hitler's Empire*. On Alwin Seifert, see Thomas Zeller, "Molding the Landscape of Nazi Environmentalism: Alwin Seifert and the Third Reich," in *How Green Were the Nazis? Nature, Environment, and Nation in the Third Reich*, ed. Franz-Josef Brüggemeier, Mark Cioc, and Thomas Zeller, 1st ed., Ohio University Press Series in Ecology and History (Athens: Ohio University Press, 2005), 147–70; Thomas Zeller, "'Ganz Deutschland sein Garten': Alwin Seifert und die Landschaft des Nationalsozialismus," in *Naturschutz und Nationalsozialismus*, ed. Joachim Radkau and Frank Uekötter (Frankfurt am Main: Campus Verlag, 2003), 273–308.

On the contested history of the forest as a German national symbol, see Jeffrey K. Wilson, *The German Forest: Nature, Identity, and the Contestation of a National Symbol, 1871–1914* (Toronto: University of Toronto Press, 2012); Michael Imort, "A Sylvan People: Wilhelmine Forestry and the Forest as a Symbol of Germandom," in *Germany's Nature: Cultural Landscapes and Environmental History*, ed. Thomas Lekan and Thomas Zeller (New Brunswick, NJ: Rutgers University Press, 2005), 81–109; Albrecht Lehmann, "Der deutsche Wald," in *Deutsche Erinnerungsorte*, ed. Etienne François and Hagen Schulze, 3 vols. (Munich: Beck, 2001), 3:187–200. On the colonial and postcolonial aspects of German forest management, see Thaddeus Sunseri, "Exploiting the Urwald: German Post-colonial Forestry in Poland and Central Africa, 1900–1960," *Past & Present* 214, no. 1 (February 1, 2012): 305–42.

8. See Gert Gröning and Joachim Wolschke-Bulmahn, *Die Liebe zur Landschaft: Der Drang nach Osten: Zur Entwicklung der Landespflege im Nationalsozialismus und während des Zweiten Weltkrieges in den "eingegliederten Ostgebieten"* (Munich: Minerva-Publikation, 1987); Wolfgang Wippermann, *Der "deutsche Drang nach Osten": Ideologie und Wirklichkeit eines politischen Schlagwortes* (Darmstadt: Wissenschaftliche Buchgesellschaft, 1981); Robert L. Nelson, ed., *Germans, Poland, and Colonial Expansion to the East: 1850 through the Present* (New York: Palgrave Macmillan, 2009).

9. On the development of the concept of *Heimat* encompassing both regional and national forms of belonging, see Celia Applegate, *A Nation of Provincials: The German Idea of Heimat* (Berkeley: University of California Press, 1990). On the history of the conservation movement in Germany, see William H. Rollins, *A Greener Vision of Home: Cultural Politics and Environmental Reform in the German Heimatschutz Movement, 1904–1918* (Ann Arbor: University of Michigan Press, 1997); Raymond H. Dominick, *The Environmental Movement in Germany: Prophets and Pioneers, 1871–1971* (Bloomington: Indiana University Press, 1992); Thomas Lekan, *Imagining the Nation in Nature: Landscape Preservation and German Identity, 1885–1945* (Cambridge, MA: Harvard University Press, 2004); Frank Uekötter, *The Green and the Brown: A History of Conservation in Nazi Germany* (Cambridge: Cambridge University Press, 2006).

10. Gregor Thum, "Ex oriente lux—ex oriente furor: Einführung," in *Traumland Osten: Deutsche Bilder vom östlichen Europa im 20. Jahrhundert*, ed. Gregor Thum (Göttingen: Vandenhoeck & Ruprecht, 2006), 8. See also Gerd Koenen, *Der Russland-Komplex: Die Deutschen und der Osten, 1900–1945* (Munich: Beck, 2005); Henry Cord Meyer, *Drang nach Osten: Fortunes of a Slogan-Concept in German-Slavic Relations, 1849–1990* (Bern: P. Lang, 1996); Joseph Partsch, *Mitteleuropa: Die Länder und Völker von den Westalpen und dem Balkan bis an den Kanal und das Kurische Haff* (Gotha: J. Perthes, 1904), 158, 196; Klaus Fehn, "'Lebensgemeinschaft von Volk und Raum': Zur nationalsozialistischen Raum- und Landschaftsplanung in den eroberten Ostgebieten," in *Naturschutz und Nationalsozialismus*, ed. Joachim Radkau and Frank Uekötter (Frankfurt am Main: Campus, 2003), 207–24; Ekkehard Klug, "Das 'asiatische' Rußland: Über die Entstehung eines europäischen Vorurteils," *Historische Zeitschrift* 245, no. 2 (October 1, 1987): 265–89; Mark Bassin, *Imperial Visions: Nationalist Imagination and Geographical Expansion in the Russian Far East, 1840–1865* (Cambridge: Cambridge University Press, 2006).

11. Vejas Gabriel Liulevicius, *The German Myth of the East: 1800 to the Present* (Oxford: Oxford University Press, 2009); Larry Wolff, *Inventing Eastern Europe: The Map of Civilization on the Mind of the Enlightenment* (Stanford, CA: Stanford University Press, 1994). The rhetoric of irreconcilable cultural differences between Russia and the rest of Europe remained powerful.

Oswald Spengler, for example, referred to a sharp dividing line between "the Russian and the occidental spirit." Cited in Wolfgang Wippermann, *Die Deutschen und der Osten: Feindbild und Traumland* (Darmstadt: Primus, 2007), 50; Kristin Leigh Kopp, *Germany's Wild East: Constructing Poland as Colonial Space* (Ann Arbor: University of Michigan Press, 2012); Vejas Gabriel Liulevicius, *War Land on the Eastern Front: Culture, National Identity and German Occupation in World War I* (Cambridge: Cambridge University Press, 2000).

12. Vejas Gabriel Liulevicius, "Der Osten als apokalyptischer Raum: Deutsche Frontwahrnehmungen im und nach dem Ersten Weltkrieg," in *Traumland Osten: Deutsche Bilder vom östlichen Europa im 20. Jahrhundert*, ed. Gregor Thum (Göttingen: Vandenhoeck & Ruprecht, 2006), 55; Joachim Wolschke-Bulmahn, "Violence as the Basis of National Socialist Landscape Planning in the 'Annexed Eastern Areas,'" in *How Green Were the Nazis? Nature, Environment, and Nation in the Third Reich*, ed. Franz-Josef Brüggemeier, Mark Cioc, and Thomas Zeller, 1st ed., Ohio University Press Series in Ecology and History (Athens: Ohio University Press, 2005), 244.

13. For the reports of Polish atrocities, see BArch R 904/790; Robert Jan van Pelt, "Bearers of Culture, Harbingers of Destruction: The Mythos of the Germans in the East," in *Art, Culture, and Media Under the Third Reich*, ed. Richard A. Etlin (Chicago: University of Chicago Press, 2002), 98–135; Gregor Thum, "Mythische Landschaften: Das Bild vom 'Deutschen Osten' und die Zäsuren des 20. Jahrhunderts," in *Traumland Osten: Deutsche Bilder vom östlichen Europa im 20. Jahrhundert*, ed. Gregor Thum (Göttingen: Vandenhoeck & Ruprecht, 2006), 194.

14. Karl Josef Kaufmann, "Der Rückgang des Deutschtums in Westpreußen zu polnischer Zeit (1569–1772)," in *Der ostdeutsche Volksboden: Aufsätze zu den Fragen des Ostens*, ed. Wilhelm Volz (Breslau: F. Hirt, 1926), 312–15; Heinrich von Treitschke, *Origins of Prussianism*, trans. Eden Paul and Cedar Paul (London: G. Allen and Unwin, 1942), 151–52.

15. See, for example, Rudolf Kötzschke, "Die deutsche Wiederbesiedelung der ostelbischen Lande," in *Der ostdeutsche Volksboden: Aufsätze zu den Fragen des Ostens*, ed. Wilhelm Volz (Breslau: F. Hirt, 1926), 155; Kaufmann, "Der Rückgang des Deutschtums," 324.

16. Joachim Wolschke-Bulmahn, "The Nationalization of Nature and the Naturalization of the German Nation: 'Teutonic' Trends in Early Twentieth-Century Landscape Design," in *Nature and Ideology: Natural Garden Design in the Twentieth Century*, ed. Joachim Wolschke-Bulmahn (Washington, DC: Dumbarton Oaks Research Library and Collection, 1997), 187–219; Wilfried Lipp, *Natur, Geschichte, Denkmal: Zur Entstehung des Denkmalbewußtseins der bürgerlichen Gesellschaft* (Frankfurt: Campus Verlag, 1987).

17. Thum, "Mythische Landschaften," 198, 202. On the land reclamation efforts during the Weimar years, see BArch R 3601/1675, 1680, 1681, 1683, 1684, 1685, 1688; Blackbourn, *Conquest of Nature*, 21–76, 93, 144–60; Eugenie Berg, *Die Kultivierung der nordwestdeutschen Hochmoore* (Oldenburg: Isensee, 2004); Rita Gudermann, "Conviction and Constraint: Hydraulic Engineers and Agricultural Amelioration Projects in Nineteenth-Century Prussia," in *Germany's Nature: Cultural Landscapes and Environmental History*, ed. Thomas M. Lekan and Thomas Zeller (New Brunswick, NJ: Rutgers University Press, 2005), 33–54; Kathryn M. Olesko, "Geopolitics & Prussian Technical Education in the Late-Eighteenth Century," *Actes d'història de la ciència i de la tècnica* 2, no. 2 (2009): 11–44; BArch R 3601/1675: Note by Th. Echtermeyer, Director of the Teaching and Research Institute of Garden Architecture in Berlin, 1927, 83.

18. BArch R 3601/1675: "Bedenkliche Folgen einer verfehlten Agrarpolitik," *Ostpreußische Zeitung*, July 25, 1926, 30; BArch R 3601/1683: Wiedenbrück to Verein zur Förderung der

Moorkultur im deutschen Reiche, May 14, 1923; BArch R 3601/1683: "Bericht über die Tätigkeit der Bremer Abteilung des Vereins zur Förderung der Moorkultur," 1929; BArch 3601/1685: "Die Deutsche Moorkultur im Reichsnährstand" (memorandum), December 1933.

19. Michael Burleigh, *Germany Turns Eastwards: A Study of Ostforschung in the Third Reich* (London: Pan Books, 2002), 21. On the history of the SdVK, see Michael Fahlbusch, *"Wo der deutsche . . . ist, ist Deutschland": Die Stiftung für deutsche Volks- und Kulturbodenforschung in Leipzig 1920–1933* (Bochum: Brockmeyer, 1994), 49–173. Other institutions involved in research on the East during the Weimar years included the Nordostdeutsche Forschungsgemeinschaft (North-East-German Research Association) and the Publikationsstelle im Geheimen Staatsarchiv (Publishing House at the Secret State Archive); Fahlbusch, *"Wo der deutsche . . . ist, ist Deutschland,"* 63–64; Fahlbusch, *Wissenschaft im Dienst der nationalsozialistischen Politik? Die „Volksdeutschen Forschungsgemeinschaften" von 1931–1945* (Baden-Baden: Nomos, 1999). On Penck and Partsch's exchange of ideas, see Albrecht Penck and Joseph Partsch, *Briefe Albrecht Pencks an Joseph Partsch*, ed. Gerhard Engelmann (Leipzig: Enzyklopädie, 1960). On Penck's academic contributions, see Ingo Schaefer, "Der Weg Albrecht Pencks nach München, zur Geographie und zur alpinen Eiszeitforschung," *Mitteilungen der Geographischen Gesellschaft in München* 74 (1989): 5–25; Norman Henniges, "'Sehen lernen': Die Exkursionen des Wiener Geographischen Instituts und die Formierung der Praxiskultur der geographischen (Feld-)Beobachtung in der Ära Albrecht Penck (1885 bis 1906)," *Mitteilungen der Österreichischen Geographischen Gesellschaft* 156 (2013): 141–70. Aside from his more famous writings on glaciers and Alpine environments, Penck was also interested in deserts; see Albrecht Penck, "Die Morphologie der Wüsten," *Geographische Zeitschrift* 15, no. 10 (1909): 545–58. On Penck's collaboration with Brückner, see John Imbrie and Katherine Palmer Imbrie, *Ice Ages: Solving the Mystery* (Cambridge, MA: Harvard University Press, 1986), 114–17; T. P. Burt et al., eds., *The History of the Study of Landforms, or, the Development of Geomorphology*, vol. 4 (Bath: Geological Society, 2008), 399–400.

20. Albrecht Penck, "Deutsches Volk und deutsche Erde," *Die Woche* 9 (1907): 179–82. On the significance of the article for German revisionism after the First World War, see Fahlbusch, *"Wo der deutsche . . . ist, ist Deutschland,"* 207–18; GStAPK Nachlass (NL) Penck, Albrecht: Untitled manuscript by Penck, 1943; Albrecht Penck, "Deutschland als geographische Gestalt," in *Deutschland. Die natürlichen Grundlagen seiner Kultur*, ed. Johannes Walther, vol. 1 (Leipzig: Quelle & Meyer, 1928), 1–10. Penck's quote is from the introduction in *Stiftung für deutsche Volks- und Kulturbodenforschung Leipzig: Die Tagungen der Jahre 1923–1929* (Langensalza: J. Beltz, 1930), ix; Hans-Dietrich Schultz, "'Ein wachsendes Volk braucht Raum': Albrecht Penck als politischer Geograph," in *1810–2010: 200 Jahre Geographie in Berlin: an der Universität zu Berlin (ab 1810), Friedrich-Wilhelms-Universität zu Berlin (ab 1828), Universität Berlin (ab 1946), Humboldt-Universität zu Berlin (ab 1949)*, ed. Bernhard Nitz, Hans-Dietrich Schultz, and Marlies Schulz (Berlin: Geographisches Institut der Humboldt Universität zu Berlin, 2010), 91–135. On Ratzel's influence on Haushofer and Penck, see Klaus Kost, *Die Einflüsse der Geopolitik auf Forschung und Theorie der politischen Geographie von ihren Anfängen bis 1945*, Bonner Geographische Abhandlungen 76 (Bonn: Ferd. Dümmler, 1988), 236–39, 266–79.

21. Albrecht Penck, "Deutscher Volks- und Kulturboden," in *Volk unter Völkern: Bücher des Deutschtums*, ed. K. C. Loesch (Breslau: Hirt, 1925), 62–72. While Penck did not invent the terms, he played a central role in propagating them in his writings and through the SdVK; see Schultz, "Albrecht Penck als politischer Geograph," 113–14; Norman Henniges, "'Naturgesetze der Kultur':

Die Wiener Geographen und die Ursprünge der 'Volks- und Kulturbodentheorie,'" *ACME: An International Journal for Critical Geographies* 14, no. 4 (December 2015): 1309–51. For an example of the incorporation of Penck's ideas into a völkisch conception of geography, see Emil Meynen, "Völkische Geographie," *Geographische Zeitschrift* 41, no. 11 (January 1, 1935): 435–41. On the influence of Penck's ideas in general, see Jürgen Zimmerer, "Im Dienste des Imperiums: Die Geographen der Berliner Universität zwischen Kolonialwissenschaften und Ostforschung," in *Universitäten und Kolonialismus*, ed. Andreas Eckert, Jahrbuch für Universitätsgeschichte 7 (Stuttgart: Steiner, 2004), 95; Mechtild Rössler, *"Wissenschaft und Lebensraum": Geographische Ostforschung im Nationalsozialismus—ein Beitrag zur Disziplingeschichte der Geographie*, vol. 8, Hamburger Beiträge zur Wissenschaftsgeschichte (Berlin: D. Reimer, 1990), 225; Willi Oberkrome, *Volksgeschichte: Methodische Innovation und völkische Ideologisierung in der deutschen Geschichtswissenschaft 1918–1945* (Göttingen: Vandenhoeck & Ruprecht, 1993), 28–29.

22. Penck had the support of the geopolitician Karl Haushofer, who mirrored Penck's claim to "German soil" in the East in his somewhat convoluted argument about the decrease in the "power of nutrition per unit of area" in the East since the Polish takeover; see Karl Haushofer, "Die geopolitische Betrachtung grenzdeutscher Probleme," in *Volk unter Völkern: Bücher des Deutschtums*, ed. K. C Loesch (Breslau: Hirt, 1925), 188–92; Penck, "Deutscher Volks- und Kulturboden," 69.

23. Burleigh, *Germany Turns Eastwards*, 22–39; Gerd Voigt, "Aufgaben und Funktion der Osteuropa-Studien in der Weimarer Republik," *Studien über die deutsche Geschichtswissenschaft* 2 (1965): 369–99; Deutsche Akademie der Wissenschaften zu Berlin, *Atlas des deutschen Lebensraumes in Mitteleuropa. Im Auftrage der Preussischen Akademie der Wissenschaften herausgegeben*, ed. Norbert Krebs (Leipzig: Bibliographisches Institut, 1937). See also Laetitia Boehm, "Langzeitvorhaben als Akademieaufgabe: Geschichtswissenschaft in Berlin und in München," in *Wissenschaft, Krieg und die Berliner Akademie der Wissenschaften*, ed. Wolfram Fischer (Berlin: Akademie Verlag, 2000), 421; Albrecht Penck, *Die Tragfähigkeit der Erde* (Leipzig: Quelle & Meyer, 1941); Wolfgang J. Mommsen, "Wissenschaft, Krieg und die Berliner Akademie der Wissenschaften," in *Die Preussische Akademie der Wissenschaften zu Berlin 1914–1945*, ed. Wolfram Fischer (Berlin: Akademie Verlag, 2000), 18–19.

24. Eduard Mühle, "Der europäische Osten in der Wahrnehmung deutscher Historiker: Das Beispiel Hermann Aubin," in *Traumland Osten: Deutsche Bilder vom östlichen Europa im 20. Jahrhundert*, ed. Gregor Thum (Göttingen: Vandenhoeck & Ruprecht, 2006), 110–37; Eduard Mühle, *Für Volk und deutschen Osten: Der Historiker Hermann Aubin und die deutsche Ostforschung* (Düsseldorf: Droste, 2005); Eduard Mühle, "'Ostforschung': Beobachtung zu Aufstieg und Niedergang eines geschichtwissenschaftlichen Paradigmas," *Zeitschrift für Ostmitteleuropa-Forschung* 46, no. 3 (1997): 317–50 On the role of historians in the mythmaking about the "German East," see Ingo Haar, *Historiker im Nationalsozialismus: Deutsche Geschichtswissenschaft und der "Volkstumskampf" im Osten* (Göttingen: Vandenhoeck & Ruprecht, 2000); Oberkrome, *Volksgeschichte*; Christoph Kleßmann, "Osteuropaforschung und Lebensraumpolitik im Dritten Reich," in *Wissenschaft im Dritten Reich*, ed. Peter Lundgreen (Frankfurt am Main: Suhrkamp, 1985), 350–83; Hans-Christian Petersen, "'Ordnung schaffen' durch Bevölkerungsverschiebung: Peter-Heinz Seraphim oder der Zusammenhang zwischen 'Bevölkerungsfragen' und Social Engineering," *Historical Social Research / Historische Sozialforschung* 31, no. 4 (118) (January 1, 2006): 282–307. On the lack of a stable definition of the "East" in Nazi planning, see Andreas Zellhuber, *"Unsere*

Verwaltung treibt einer Katastrophe zu . . .": Das Reichsministerium für die besetzten Ostgebiete und die deutsche Besatzungsherrschaft in der Sowjetunion 1941–1945 (Munich: Vögel, 2006), 2.

25. Gerd Koenen, "Der deutsche Russland-Komplex," in *Traumland Osten: Deutsche Bilder vom östlichen Europa im 20. Jahrhundert,* ed. Gregor Thum (Göttingen: Vandenhoeck & Ruprecht, 2006), 40–41; Bassin, "Blood or Soil?"; Mark Bassin, "Race Contra Space: The Conflict between German Geopolitik and National Socialism," *Political Geography Quarterly* 6, no. 2 (April 1987): 115–34.

26. Michael A. Hartenstein, *Neue Dorflandschaften: Nationalsozialistische Siedlungsplanung in den "eingegliederten Ostgebieten" 1939 bis 1944,* Wissenschaftliche Schriftenreihe Geschichte 6 (Berlin: Köster, 1998), 25–28; Gesine Gerhard, "Breeding Pigs and People for the Third Reich: Richard Walther Darré's Agrarian Ideology," in *How Green Were the Nazis? Nature, Environment, and Nation in the Third Reich,* ed. Franz-Josef Brüggemeier, Mark Cioc, and Thomas Zeller (Athens: Ohio University Press, 2005), 129–46; Richard Walther Darré, *Neuadel aus Blut und Boden* (Munich: J. F. Lehmann, 1930), 84–85; Koehl, *RKFDV,* 27–28.

27. Marcel Herzberg, *Raumordnung im nationalsozialistischen Deutschland* (Dortmund: Dortmunder Vertrieb für Bau- und Planungsliteratur, 1997), 108–11; Burleigh, *Germany Turns Eastwards,* 144; Gröning and Wolschke-Bulmahn, *Der Drang nach Osten,* 194. On the eclectic spheres of influence of the RKF, see Götz Aly and Susanne Heim, *Vordenker der Vernichtung: Auschwitz und die deutschen Pläne für eine neue europäische Ordnung* (Hamburg: Hoffmann und Campe, 1991), 125–31. On Wiepking's position in the RKF, see Rössler, *Wissenschaft und Lebensraum,* 8:167. After the war, Wiepking denied ever having been a "special deputy" of the RKF and claimed to have worked merely as an advisor. His central role in the drafting of the Landscape Decree (see below) makes that claim seem highly dubious. See NLSO Dep 72/39 and Kellner, "Heinrich Friedrich Wiepking," 322.

28. See Karl Heinz Roth, "'Generalplan Ost'–'Gesamtplan Ost': Forschungsstand, Quellenprobleme, neue Ergebnisse," in *Der "Generalplan Ost": Hauptlinien der nationalsozialistischen Planungs- und Vernichtungspolitik,* ed. Mechtild Rössler and Sabine Schleiermacher (Berlin: Akademie Verlag, 1993), 69; Mechtild Rössler and Sabine Schleiermacher, "Der 'Generalplan Ost' und die 'Modernität' der Großraumordnung. Eine Einführung," in *Der "Generalplan Ost": Hauptlinien der nationalsozialistischen Planungs- und Vernichtungspolitik,* ed. Mechtild Rössler and Sabine Schleiermacher (Berlin: Akademie Verlag, 1993), 9; Willi Oberkrome, *Ordnung und Autarkie: Die Geschichte der deutschen Landbauforschung, Agrarökonomie und ländlichen Sozialwissenschaft im Spiegel von Forschungsdienst und DFG (1920–1970)* (Stuttgart: Steiner, 2009), 104–14; Ariane Leendertz, *Ordnung schaffen: Deutsche Raumplanung im 20. Jahrhundert* (Göttingen: Wallstein, 2008), 148–53; Gröning and Wolschke-Bulmahn, *Der Drang nach Osten,* 25; Heinrich Wiepking-Jürgensmann, *Die Landschaftsfibel* (Berlin: Deutsche Landbuchhandlung, 1942), 24; J. O. Plassmann, "Deutsche Landgestaltung in völkischer Schau," in *Landvolk im Werden: Material zum ländlichen Aufbau in den neuen Ostgebieten und zur Gestaltung des dörflichen Lebens,* ed. Konrad Meyer (Berlin: Deutsche Landbuchhandlung, 1941), 271; Artur Schürmann, "Festigung deutschen Volkstums in den eingegliederten Ostgebieten," *Reich, Volksordnung, Lebensraum* 6 (1944): 475–538.

29. Christof Mauch, *Nature in German History* (New York: Berghahn Books, 2004), 86–90. Blackbourn has described the focus of Nazi planners on the Pripet Marshes in today's Ukraine and Belarus; see Blackbourn, *Conquest of Nature,* 251–78; H. Kuron, "Die Bodenerosion in Europa: Eine zusammenfassende Darstellung ohne die Gebirge," *Der Forschungsdienst* 16, no. 1 (1943): 15.

30. Paul Sokolowski, *Die Versandung Europas: ... eine andere, große russische Gefahr* (Berlin: Deutsche Rundschau, 1929), 102; David Blackbourn, "'The Garden of Our Hearts': Landscape, Nature, and Local Identity in the German East," in *Localism, Landscape, and the Ambiguities of Place: German-Speaking Central Europe, 1860–1930*, ed. James N. Retallack and David Blackbourn (Toronto: University of Toronto Press, 2007), 158.

31. Werner Junge, "Aufbauelemente einer deutschen Heimatlandschaft," in *Landvolk im Werden: Material zum ländlichen Aufbau in den neuen Ostgebieten und zur Gestaltung des dörflichen Lebens*, ed. Konrad Meyer (Berlin: Deutsche Landbuchhandlung, 1941), 303–10.

32. Guido Görres, "Gestaltungsaufgaben im neuen Ostpreußen," *Neues Bauerntum* 42 (1940): 245, emphasis added.

33. Karl Schlögel, "Die russische Obsession: Edwin Erich Dwinger," in *Traumland Osten: Deutsche Bilder vom östlichen Europa im 20. Jahrhundert*, ed. Gregor Thum (Göttingen: Vandenhoeck & Ruprecht, 2006), 66–87; Karl Christian Thalheim and Arnold Hillen Ziegfeld, eds., *Der deutsche Osten, seine Geschichte, sein Wesen und seine Aufgabe* (Berlin: Propyläen-Verlag, 1936). It is interesting to note that Russian articles had already furthered the image of a particular German quality of soil care in the late nineteenth century by praising German Mennonites for having transformed arid steppe lands well-ordered agricultural lands; see David Moon, *The Plough That Broke the Steppes: Agriculture and Environment on Russia's Grasslands, 1700–1914* (Oxford: Oxford University Press, 2013), 278.

34. Interestingly, Humboldt already foreshadowed the negative connotations connected to the steppe by referring to the "dreary steppe-lands of Northern Asia" in his main work; see Alexander von Humboldt, *Kosmos: A General Survey of the Physical Phenomena of the Universe*, vol. 1 (London: H. Baillière, 1845), ix; Alexander von Humboldt, *Ansichten der Natur mit wissenschaftlichen Erläuterungen* (Tübingen: J. G. Cotta, 1808); Eduard A. Rübel, "Heath and Steppe, Macchia and Garigue," *Journal of Ecology* 2 (1914): 233–34. On the history of research into steppe environments in Central Asia, see Moon, *The Plough That Broke the Steppes*; David Moon, "The Steppe as Fertile Ground for Innovation in Conceptualizing Human-Nature Relationships," *Slavonic and East European Review* 93, no. 1 (January 1, 2015): 16–38.

35. Robert Gradmann, *Das Pflanzenleben der Schwäbischen Alb: Mit Berücksichtigung der angrenzenden Gebiete Süddeutschlands*, 2 vols. (Tübingen: Verlag des Schwäbischen Albvereins, 1898), vol. 1; see also A. Nehring, "Die Ursachen der Steppenbildung in Europa," *Geographische Zeitschrift* 1 (1895): 152–63; Johannes Walther, "Der Begriff der Steppe," *Petermanns Mitteilungen aus Justus Perthes' Geographischer Anstalt* 65 (1919): 102; Wladimir Köppen, "Klassifikation der Klimate nach Temperatur, Niederschlag and Jahreslauf," *Petermanns geographische Mitteilungen* 64 (1918): 193–203.

36. Robert Gradmann, "Zur prähistorischen Siedlungsgeographie des norddeutschen Tieflands," in *Festgabe der Philosophischen Fakultät der Friedrich-Alexander-Universität Erlangen zur 55. Versammlung deutscher Philologen und Schulmänner* (Erlangen: K. Döres, 1925), 1–10; Robert Gradmann, *Volkstum und Rasse in Süddeutschland: Rede beim Antritt des Rektorates der Bayerischen Friedrich-Alexanders-Universität Erlangen am 4. November 1925* (Erlangen: K. Döres, 1926); Robert Gradmann, "Zur deutschen Rassenkunde," *Monatsschrift für akademisches Leben: Fränkische Hochschulzeitung*, no. 5 (February 1928): 105–6; *Stiftung für deutsche Volks- und Kulturbodenforschung Leipzig*, 49, 175–77.

37. Robert Gradmann, *Die Steppen des Morgenlandes in ihrer Bedeutung für die Geschichte der menschlichen Gesittung* (Stuttgart: J. Engelhorns Nachfolger, 1934), 15; Friedrich Ratzel, *Der*

Lebensraum: Eine biogeographische Studie (Tübingen: H. Laupp, 1901), 29; see also Hans Mortensen, "Probleme der deutschen morphologischen Wüstenforschung," *Naturwissenschaften* 18, no. 28 (July 1, 1930): 629–37.

38. Alwin Seifert, *Im Zeitalter des Lebendigen. Natur, Heimat, Technik,* 3rd ed. (Planegg: Müller, 1943); Blackbourn, *Conquest of Nature,* 285–93. On the history of the concept of *Versteppung* and its use in Germany, see Axel Zutz, "Fear of the 'Steppes': Soil Protection and Landscape Planning in Germany 1930–1960 between Politics and Science," *Global Environment* 8, no. 2 (January 1, 2015): 380–409. As Zutz shows, the debates about *Versteppung* did not occur in a vacuum, but were connected to wider scientific and cultural trends. Lekan, *Imagining the Nation in Nature,* 212–51; Franz Wilhelm Seidler, *Fritz Todt: Baumeister des Dritten Reiches,* 2nd ed. (Frankfurt am Main: Ullstein, 1988); Karsten Runge, *Entwicklungstendenzen der Landschaftsplanung: Vom frühen Naturschutz bis zur ökologisch nachhaltigen Flächennutzung* (Berlin: Springer, 1998), 27–29; Zeller, "Molding the Landscape of Nazi Environmentalism," 156–57. Seifert was close to the Anthroposophical movement, but not a member of the Anthroposophic Society; see Uwe Werner, *Anthroposophen in der Zeit des Nationalsozialismus (1933–1945)* (Munich: Oldenbourg, 1999), 87–89.

39. Seifert's article was first published as Alwin Seifert, "Die Versteppung Deutschlands," *Deutsche Technik* 4 (1936): 423–27, 490–92. On the central role of this journal in disseminating National Socialist ideas and ideologies of technology, see Helmut Maier, "Nationalsozialistische Technikideologie und die Politisierung des 'Technikerstandes': Fritz Todt und die Zeitung 'Deutsche Technik,'" in *Technische Intelligenz und "Kulturfaktor Technik": Kulturvorstellungen von Technikern und Ingenieuren zwischen Kaiserreich und früher Bundesrepublik Deutschland,* ed. Burkhard Dietz, Michael Fessner, and Helmut Maier (Münster: Waxmann, 1996), 253–68; Karl-Heinz Bernhardt, "Alexander von Humboldts Beitrag zu Entwicklung und Institutionalisierung von Meteorologie und Klimatologie im 19. Jahrhundert," *Algorismus* no. 41 (2003): 213–14; Alwin Seifert, "Gedanken über bodenständige Gartenkunst," *Gartenkunst* 42 (1929): 118–23, 131–32, 175–78, 191–95; ADM NL 133/007: Seifert, "Bodenständige Gartenkunst, Vortrag gehalten im bayrischen Landesvereins für Heimatschutz," April 11, 1930, n.p. See also Zeller, "Molding the Landscape of Nazi Environmentalism," 154–55; Gert Gröning and Joachim Wolschke-Bulmahn, "The National Socialist Garden and Landscape Ideal: Bodenständigkeit (Rootedness in the Soil)," in *Art, Culture, and Media under the Third Reich,* ed. Richard A Etlin (Chicago: University of Chicago Press, 2002), 73–97; Otto Jaekel, *Die Gefahren der Entwässerung unseres Landes,* Mitteilungen aus dem geologisch-palaeontogischen Institut der Universität Greifswald 4 (Greifswald: L. Bamberg, 1922).

40. Paul Kessler, "Einige Wüstenerscheinungen aus nicht aridem Klima," *Geologische Rundschau* 4, no. 7 (November 1, 1913): 413–23; Paul Kessler, *Das Klima der jüngsten geologischen Zeiten und die Frage einer Klimaänderung in der Jetztzeit* (Stuttgart: Schweizerbart, 1923); Richard Scherhag, "Eine bemerkenswerte Klimaänderung über Nordeuropa," *Annalen der Hydrographie und maritimen Meteorologie* 64 (March 1936): 96–100; Richard Scherhag, "Die gegenwärtige Milderung der Winter und ihre Ursachen," *Annalen der Hydrographie und maritimen Meteorologie* 67 (June 1939): 292–303.

41. See Zutz, "Fear of the 'Steppes'"; Paul B. Sears, *Deserts on the March* (Norman: University of Oklahoma Press, 1935); Joachim Radkau, *The Age of Ecology: A Global History,* trans. Patrick Camiller (Cambridge: Polity Press, 2014), 46–60. On the international dimension of the Dust Bowl, see Sarah T. Phillips, "Lessons from the Dust Bowl: Dryland Agriculture and Soil Erosion

in the United States and South Africa, 1900–1950," *Environmental History* 4, no. 2 (April 1, 1999): 245–66; Sabine Sauter, "Australia's Dust Bowl: Transnational Influences in Soil Conservation and the Spread of Ecological Thought," *Australian Journal of Politics & History* 61, no. 3 (September 1, 2015): 352–65; Hannah Holleman, "De-naturalizing Ecological Disaster: Colonialism, Racism and the Global Dust Bowl of the 1930s," *Journal of Peasant Studies* 44, no. 1 (January 2, 2017): 234–60. For the classic study on the Dust Bowl, see Donald Worster, *Dust Bowl: The Southern Plains in the 1930s* (Oxford: Oxford University Press, 2004).

42. ADM NL 133/013: Seifert, "Hat der Wald Einfluss auf das Klima?," n.d. (probably 1944 or 1945), 1; Alwin Seifert, "Die Versteppung Deutschlands," in *Im Zeitalter des Lebendigen. Natur, Heimat, Technik* (Planegg: Müller, 1943), 29–36.

43. Alwin Seifert, "Naturnahe Wasserwirtschaft," in *Im Zeitalter des Lebendigen. Natur, Heimat, Technik* (Planegg: Müller, 1943), 66; BArch NS 26/1188: Seifert to Todt, September 6, 1941; Alwin Seifert, "Mahnung an die Burgherren," *Deutsche Technik* 9 (January 1941): 9–13; ADM NL 133/010: Seifert, "Die Gefährdung der Lebensgrundlagen des Dritten Reiches durch die heutigen Arbeitsweisen des Kultur- u. Wasserbaus," August 13, 1935, 6.

44. See, for example, W. Koehne, "Zur Frage der 'Versteppung,'" *Deutsche Wasserwirtschaft: Zentralblatt für Wasserbau, Wasserkraft und Wasserwirtschaft* 32, no. 2 (February 1, 1937): 33–36; J. Buck, "Landeskultur und Natur," *Deutsche Landeskultur-Zeitung* 2 (1937): 48–54; Riecke, "Erörterung über die Gefahren einer Versteppung Deutschlands," *Deutsche Landeskultur-Zeitung* 7, no. 3 (1938): 112–14. Darré's quote is cited in Zeller, "Molding the Landscape of Nazi Environmentalism," 156; ADM NL 133/021: Seifert to H. Löber, November 23, 1937; Fritz Todt, ed., *Die Versteppung Deutschlands? (Kulturwasserbau und Heimatschutz)* (Berlin: Weicher, 1938); Seidler, *Fritz Todt*, 279–80; O. Uhden, "Die Unhaltbarkeit der Seifert'schen Versteppungtheorie," *Deutsche Landeskultur-Zeitung* 10, no. 9 (September 1, 1941): 177–82; ADM NL 133/008: Seifert, "Versteppung, Windschutz und kein Ende," March 16, 1949, n.p.

45. Gerhard Lenz, "Ideologisierung und Industrialisierung der Landschaft im Nationalsozialismus am Beispiel des Großraumes Bitterfeld-Dessau," in *Veränderung der Kulturlandschaft: Nutzungen—Sichtweisen—Planungen*, ed. Günter Bayerl and Torsten Meyer (Münster: Waxmann, 2003), 177–97; ADM NL 133/023: Seifert to Friedrich Reck, February 8, 1944, n.p.; Alwin Seifert, "Die Zukunft der ostdeutschen Landschaft," *Bauen, Siedeln, Wohnen* 20, no. 9 (1940): 312–16.

Chapter 7

1. The term "bloodlands" is borrowed from Timothy Snyder, *Bloodlands: Europe Between Hitler and Stalin* (New York: Basic Books, 2010). On the development of German environmental chauvinism in the colonies, see William H. Rollins, "Imperial Shades of Green: Conservation and Environmental Chauvinism in the German Colonial Project," *German Studies Review* 22, no. 2 (May 1, 1999): 187–213.

2. Martin Broszat, *Nationalsozialistische Polenpolitik 1939–1945*, Schriftenreihe der Vierteljahreshefte für Zeitgeschichte 2 (Stuttgart: Deutsche-Verlags-Anstalt, 1961), 85; Wilhelm Zoch, *Neuordnung im Osten: Bauernpolitik als deutsche Aufgabe* (Berlin: Deutsche Landbuchhandlung, 1940), 153.

3. Thomas Zeller, "Molding the Landscape of Nazi Environmentalism: Alwin Seifert and the Third Reich," in *How Green Were the Nazis? Nature, Environment, and Nation in the Third*

Reich, ed. Franz-Josef Brüggemeier, Mark Cioc, and Thomas Zeller, 1st ed., Ohio University Press Series in Ecology and History (Athens: Ohio University Press, 2005), 158; Heinrich Wiepking-Jürgensmann, *Das Haus in der Landschaft* (Berlin: Gartenschönheit, 1927); Emanuel Hübner, "Haus Schandau. Ein Mannschaftsgebäude des Olympischen Dorfes von 1936," in *NS-Architektur: Macht und Symbolpolitik*, ed. Tilman Harlander and Wolfram Pyta (Berlin: Lit, 2010), 101–18; NLSO Dep 72b/135: Wiepking, "Wie betrachtet der Landschafts- und Gartengestalter das Reichsnaturschutzgesetz?" (unpublished manuscript), n.d. (ca. 1937–38), n.p. On the decrease of agricultural land in the Third Reich because of new military areas and the construction of the highway system, see Willi Oberkrome, *Ordnung und Autarkie: Die Geschichte der deutschen Landbauforschung, Agrarökonomie und ländlichen Sozialwissenschaft im Spiegel von Forschungsdienst und DFG (1920–1970)* (Stuttgart: Steiner, 2009), 170–73; Heinrich Wiepking-Jürgensmann, "Der deutsche Osten: Eine vordringliche Aufgabe für unsere Studierenden," *Die Gartenkunst* 52 (1939): 193. See also Birgit Karrasch, "Die 'Gartenkunst' im Dritten Reich," *Garten und Landschaft* 100 (June 1990): 52–56; BArch R 49/511: "Vereinbarung zwischen dem RKF und dem RFA," March 20, 1942, n.p.; BArch R 49/898: Lutz Heck and Konrad Meyer, "Vereinbarung zwischen dem RKF-Stabshauptamt und dem Reichsforstmeister als Oberster Naturschutzbehörde," March 20, 1942, n.p.

4. Heinrich Wiepking-Jürgensmann, "Gegen den Steppengeist," *Das Schwarze Korps*, October 15, 1942, 4.

5. See Thomas Zeller, "'Ich habe die Juden möglichst gemieden': Ein aufschlußreicher Briefwechsel zwischen Heinrich Wiepking und Alwin Seifert," *Garten und Landschaft* 105, no. 8 (1995): 4–5; Gert Gröning and Joachim Wolschke-Bulmahn, *Die Liebe zur Landschaft: Der Drang nach Osten: Zur Entwicklung der Landespflege im Nationalsozialismus und während des Zweiten Weltkrieges in den "eingegliederten Ostgebieten"* (Munich: Minerva-Publikation, 1987), 196; NLSO Dep 72b/20: Wiepking to Seifert, September 16, 1939; Seifert to Wiepking, September 24, 1939; Wilhelm Hübotter (landscape architect) to Wiepking, October 12, 1939; Seifert to Joseph Pertl (Head of the Municipal Parks Department in Berlin), October 19, 1939; Pertl to Wiepking, October 31, 1939, n.p.; Thomas Zeller, *Driving Germany: The Landscape of the German Autobahn, 1930–1970* (New York: Berghahn Books, 2007), 92; NLSO Dep 72b/124: Wiepking-Jürgensmann, Remarks about Seifert's article "Die Heckenlandschaft" (July 1942), September 24, 1942, n.p.

6. NLSO Dep 72b/116: Wiepking, "Wasser und Wasserbau in der Landschaft," January 24, 1938, n.p.; Alwin Seifert, "Die Versteppung Deutschlands," in *Im Zeitalter des Lebendigen. Natur, Heimat, Technik* (Planegg: Müller, 1943), 40; Heinrich Wiepking-Jürgensmann, "Die Erhaltung der Schöpferkraft," *Das Schwarze Korps*, September 17, 1942, 4; Heinrich Wiepking-Jürgensmann, "Die Landschaft der Deutschen," *Das Schwarze Korps*, October 8, 1942, 4; Heinrich Wiepking-Jürgensmann, "Aufgaben und Ziele deutscher Landschaftspolitik," *Die Gartenkunst* 53 (1940): 84; NLSO Dep 72b/11: Handwritten note by Wiepking, n.d., n.p.; Heinrich Wiepking-Jürgensmann, "Der deutsche Mensch in seiner Beziehung zum Baum und zum Walde," *Raumforschung und Raumordnung*, no. 2 (1938): 544; Heinrich Wiepking-Jürgensmann, "Deutsche Landschaft als deutsche Ostaufgabe," *Neues Bauerntum* 32, no. 4/5 (1940): 133–34; Heinrich Wiepking-Jürgensmann, "Das Landschaftsgesetz des weiten Ostens," *Neues Bauerntum* 34, no. 1 (1942): 5–18.

7. Alwin Seifert, "Natur und Technik im deutschen Straßenbau," in *Im Zeitalter des Lebendigen. Natur, Heimat, Technik* (Planegg: Müller, 1943), 12; BArch R 49/898: Seifert, "Vorläufige

Richtlinien für die Bepflanzung der Reichs- und Landstraßen in den neu eingegeliederten Ostgebieten," February 10, 1942, n.p.

8. Heinrich Wiepking-Jürgensmann, "Die Landesverschönerungskunst im Wandel der letzten 150 Jahre," *Zentralblatt der Bauverwaltung vereinigt mit Zeitschrift für Bauwesen* 60, no. 22
(1940): 320; Heinrich Wiepking-Jürgensmann, "Das Grün im Dorf und in der Feldmark," *Bauen,
Siedeln, Wohnen* 1.7, no. 18 (1940): 445; NLSO Dep 72b/124: Wiepking, "Der deutsche Weich-
selraum," October 23, 1939, n.p.

9. BArch R 49/511: Wiepking to Himmler, n.d. (probably 1941), n.p. On an argument for
Wiepking as part of the nature conservation movement, see Frank Uekötter, *The Green and the
Brown: A History of Conservation in Nazi Germany* (Cambridge: Cambridge University Press,
2006), 156–57; BArch R 49/513: Wiepking to Himmler, "Bericht über die Teilnahme an der
Tagung der Reichsstiftung für deutsche Ostforschung. Arbeitskreis für die Wiederbewaldung
des Ostens am 28. und 29. Januar 1942," n.p. On the "reconciliation of nature and technology"
under the Nazis, see Thomas Rohkrämer, *A Single Communal Faith? The German Right from
Conservatism to National Socialism* (New York: Berghahn Books, 2007), 230–33; NLSO Dep
72b/126: Wiepking, "Zum Thema 'Bäuerliche Leistung'" (manuscript), June 30, 1941, n.p.

10. NLSO Dep 72b/7: Wiepking to Kube (Head of the Municipal Parks Department in Hannover), September 27, 1934, n.p.; NLSO Dep 72b/134: Wiepking, "Gartenheimat des Volkes:
Bilder und Gedanken (unpublished manuscript), n.d., n.p.; NLSO Dep 72b/134: Wiepking, "Gartenheimat des Volkes: Bilder und Gedanken (unpublished manuscript), n.d., n.p.; see also Heinrich Wiepking-Jürgensmann, "Um die Erhaltung der Kulturlandschaft," *Raumforschung und Raumordnung* 2, no. 3 (1938): 122; BArch R 49/165: Wiepking, "Entwurf 2, Allgemeine Anordnung
über die Gestaltung der Landschaft in den eingegliederten Ostgebieten," 42; NLSO Dep 72b/212:
Wiepking, "Über die Gefahren der Sandverwehung," n.d. (ca. 1941), n.p.; NLSO Dep 72b/116:
Wiepking, Note on the "Difference between East and West," 1942 n.p.; Heinrich Wiepking-
Jürgensmann, *Die Landschaftsfibel* (Berlin: Deutsche Landbuchhandlung, 1942), 282–83.

11. Wiepking-Jürgensmann, "Der deutsche Osten," 193; Heinrich Wiepking-Jürgensmann,
"Wie muß eine gesunde Landschaft aussehen?," *Neues Bauerntum* 33 (1941): 162–66. On the rise
of medical imagery in the German conservation movement in the interwar period, see Thomas
Lekan, *Imagining the Nation in Nature: Landscape Preservation and German Identity, 1885–1945*
(Cambridge, MA: Harvard University Press, 2004), 92–94. On earlier ideas of the salubrious
effect of forests, see Jeffrey K. Wilson, *The German Forest: Nature, Identity, and the Contestation
of a National Symbol, 1871–1914* (Toronto: University of Toronto Press, 2012), 66–69; Wiepking-
Jürgensmann, *Die Landschaftsfibel*, 13, 308; The *Landschaftsfibel's* importance for planning is
underlined by Mäding's statement that its contents were to be regarded as "the law" by all officials working in the new settlement areas; see BArch R 49/513: Mäding, "Vermerk über die
Rechtslage auf dem Sachgebiet Landschaftsgestaltung," September 27, 1941, n.p.

12. Erwin Aichinger, "Pflanzen- und Menschengesellschaft, ein biologischer Vergleich,"
Biologia Generalis 17 (1943): 56–79; James Rodger Fleming, *Historical Perspectives on Climate
Change* (New York: Oxford University Press, 1998), 11–20; Jan Golinski, *British Weather and
the Climate of Enlightenment* (Chicago: University of Chicago Press, 2007), 137–202;
Charles W. J. Withers, *Placing the Enlightenment: Thinking Geographically about the Age of Reason* (Chicago: University of Chicago Press, 2007), 129–63; Theodore Feldman, "Late Enlightenment Meteorology," in *The Quantifying Spirit in the 18th Century*, ed. Tore Frängsmyr, H. L.

Heilborn, and Robin E. Rider (Berkeley: University of California Press, 1990), 143–78; Willy Hellpach, *Die geopsychischen Erscheinungen: Wetter und Klima und Landschaft in ihrem Einfluss auf das Seelenleben* (Leipzig: W. Engelmann, 1911); Willy Hellpach, *Einführung in die Völkerpsychologie* (Stuttgart: F. Enke, 1938); Willy Hellpach, *Geopsyche: Die Menschenseele unterm Einfluß von Wetter und Klima, Boden und Landschaft,* 5th ed. (Leipzig: Wilhelm Engelmann, 1939). On Hellpach's climatological ideas, see Gröning and Wolschke-Bulmahn, *Der Drang nach Osten,* 126; Nico Stehr, "The Ubiquity of Nature: Climate and Culture," *Journal of the History of the Behavioral Sciences* 32, no. 2 (1996): 151–59. For a biography of Hellpach, see Claudia-Anja Kaune, *Willy Hellpach (1877–1955): Biographie eines liberalen Politikers der Weimarer Republik* (Frankfurt am Main: P. Lang, 2005).

13. See Broszat, *Nationalsozialistische Polenpolitik,* 98. On the general enthusiasm for planning in the Third Reich, see Ariane Leendertz, *Ordnung schaffen: Deutsche Raumplanung im 20. Jahrhundert* (Göttingen: Wallstein, 2008), 107–216; Lekan, *Imagining the Nation in Nature,* 215–51; Oberkrome, *Ordnung und Autarkie,* 212–25; Klaus Fehn, "Die Auswirkungen der Veränderungen der Ostgrenze des Deutschen Reiches auf das Raumordnungskonzept des NS-Regimes (1938–1942)," *Siedlungsforschung: Archäologie, Geschichte, Geographie* 9 (1991): 199–227; Mechtild Rössler and Sabine Schleiermacher, "Der 'Generalplan Ost' und die 'Modernität' der Großraumordnung. Eine Einführung," in *Der "Generalplan Ost": Hauptlinien der nationalsozialistischen Planungs- und Vernichtungspolitik,* ed. Mechtild Rössler and Sabine Schleiermacher (Berlin: Akademie Verlag, 1993), 7–11; Bruno Wasser, *Himmlers Raumplanung im Osten: Der Generalplan Ost in Polen, 1940–1944* (Basel: Birkhäuser, 1993); Götz Aly and Susanne Heim, *Vordenker der Vernichtung: Auschwitz und die deutschen Pläne für eine neue europäische Ordnung* (Hamburg: Hoffmann und Campe, 1991), 394–440; Klaus Fehn, "'Lebensgemeinschaft von Volk und Raum': Zur nationalsozialistischen Raum- und Landschaftsplanung in den eroberten Ostgebieten," in *Naturschutz und Nationalsozialismus,* ed. Joachim Radkau and Frank Uekötter (Frankfurt am Main: Campus, 2003), 207–24; Leendertz, *Ordnung schaffen,* 143–86; Tiago Saraiva, *Fascist Pigs: Technoscientific Organisms and the History of Fascism* (Cambridge, MA: MIT Press, 2018), 186ff. For a collection of primary sources on the *Generalplan,* see Czesław Madajczyk, ed., *Vom Generalplan Ost zum Generalsiedlungsplan,* Einzelveröffentlichungen der Historischen Kommission zu Berlin 80 (Munich: Saur, 1994); Karl Heinz Roth, "'Generalplan Ost'—'Gesamtplan Ost': Forschungsstand, Quellenprobleme, neue Ergebnisse," in *Der "Generalplan Ost": Hauptlinien der nationalsozialistischen Planungs- und Vernichtungspolitik,* ed. Mechtild Rössler and Sabine Schleiermacher (Berlin: Akademie Verlag, 1993), 25–95. Patrick Bernhard has also argued that the *Generalplan's* settlement schemes were partly modeled on Italian designs for colonial Libya: Bernhard, "Hitler's Africa in the East: Italian Colonialism as a Model for German Planning in Eastern Europe," *Journal of Contemporary History* 51, no. 1 (2016): 61–90.

14. BArch R 49/230: "Grundlagen der Planung," n.d. (probably 1939), n.p.; BArch R 49/157: "Planungsgrundlagen für den Aufbau der Ostgebiete," n.d. (probably 1940), 10. See also Helmut Heiber, "Der Generalplan Ost: Vorbemerkung," *Vierteljahrshefte für Zeitgeschichte* 6, no. 3 (July 1, 1958): 285. For a short chronology of the *Generalplan,* see Robert Gellately, "Review of Vom Generalplan Ost Zum Generalsiedlungsplan by Czeslaw Madajczyk; Der 'Generalplan Ost.' Hauptlinien Der Nationalsozialistischen Planungs- Und Vernichtungspolitik by Mechtild Rössler; Sabine Schleiermacher," *Central European History* 29, no. 2 (January 1, 1996): 270–74; Rössler and Schleiermacher, "Der 'Generalplan Ost' und die 'Modernität' der

Großraumordnung. Eine Einführung," 7; BArch NS 19/1739: Meyer to Himmler, July 15, 1941, 2; Madajczyk, *Vom Generalplan Ost zum Generalsiedlungsplan*, viii–x. For simplicity's sake, I refer to the entirety of the plans as the *Generalplan* in this chapter; Konrad Meyer, "Neues Landvolk," in *Landvolk im Werden: Material zum ländlichen Aufbau in den neuen Ostgebieten und zur Gestaltung des dörflichen Lebens*, ed. Konrad Meyer (Berlin: Deutsche Landbuchhandlung, 1941), 22; Karl-Heinz Ludwig, *Technik und Ingenieure im Dritten Reich* (Düsseldorf: Droste, 1974), 426–31; André Mineau, *SS Thinking and the Holocaust* (Amsterdam: Rodopi, 2012), 31.

15. Konrad Meyer, "Planung und Ostaufbau," *Raumforschung und Raumordnung* 5, no. 9 (1941): 392. See also Adalbert Forstreuter, *Deutsches Ringen um den Osten: Kampf und Anteil der Stämme und Gaue des Reiches*, ed. Rudolf Jung (Berlin: C. A. Weller, 1940), 362; Fritz Wächtler, ed., *Reichsaufbau im Osten* (Munich: Deutscher Volksverlag, 1941), 34; Fehn, "'Lebensgemeinschaft von Volk und Raum,'" 216–17; BArch R 40/157a: "Generalplan Ost, rechtliche, wirtschaftliche und räumliche Grundlagen des Ostaufbaus, vorgelegt von Konrad Meyer," June 1942, 54. On the planned transformations of cities and villages in the annexed areas in the East, see Michael A. Hartenstein, *Neue Dorflandschaften: Nationalsozialistische Siedlungsplanung in den "eingegliederten Ostgebieten" 1939 bis 1944*, Wissenschaftliche Schriftenreihe Geschichte 6 (Berlin: Köster, 1998); Niels Gutschow, *Ordnungswahn: Architekten planen im "eingedeutschten Osten,"* 1939–1945 (Gütersloh and Berlin: Bertelsmann Fachzeitschriften and Birkhäuser, 2001); Reichsführer-SS (Heinrich Himmler), ed., *Der Untermensch* (Berlin: Nordland Verlag, 1942), n.p.; Dietrich Eichholtz, "Der 'Generalplan Ost' als genozidale Variante der imperialistischen Ostexpansion," in *Der "Generalplan Ost": Hauptlinien der nationalsozialistischen Planungs- und Vernichtungspolitik*, ed. Mechtild Rössler and Sabine Schleiermacher (Berlin: Akademie Verlag, 1993), 118–30. On slave labor during the war, see Adam Tooze, *The Wages of Destruction: The Making and Breaking of the Nazi Economy* (New York: Penguin, 2008), 524–38.

16. Wilhelm Zoch, "Neue Ordnung im Osten," *Neues Bauerntum* 32, no. 3 (1940): 85; Heinrich Wiepking-Jürgensmann, "Dorfbau und Landschaftsgestaltung," in *Neue Dorflandschaften: Gedanken und Pläne zum ländlichen Aufbau in den neuen Ostgebieten und im Altreich*, ed. Stabshauptamt des Reichskommissars für die Festigung Deutschen Volkstums (Berlin: Sohnrey, 1943), 25.

17. Heinrich Wiepking-Jürgensmann, "Raumordnung und Landschaftsgestaltung: Um die Erhaltung der schöpferischen Kräfte des deutschen Volkes," *Raumforschung und Raumordnung* 5, no. 1 (1941): 19. On Wiepking's idea of climate moderation through bodies of water, see Wiepking-Jürgensmann, *Die Landschaftsfibel*, 137, 202ff.; Wiepking-Jürgensmann, "Gegen den Steppengeist," 4; Wiepking-Jürgensmann, "Deutsche Landschaft als deutsche Ostaufgabe," 133. On the "special position" of technology in the occupied areas, see Ludwig, *Technik und Ingenieure im Dritten Reich*, 431–36. See also Paul Josephson, "Technology and Politics in Totalitarian Regimes: Nazi Germany," in *Death by Design: Science, Technology, and Engineering in Nazi Germany*, ed. Eric Katz (New York: Pearson Longman, 2006), 71–86.

18. See, e.g., BArch R 73/12846: Mäding to Meyer, "Betr. Forschungsmittel," April 3, 1943; Wilhelm Kreutz, "Methoden der Klimasteuerung: Praktische Wege in Deutschland und der Ukraine," *Der Forschungsdienst* 15, no. 5 (1943): 256–81; see also Gröning and Wolschke-Bulmahn, *Der Drang nach Osten*, 122–23; BArch R 3601/3106a: Report by E. Gödecke about climate amelioration, n.d., 76–80.

19. Erhard Mäding, *Landespflege: Die Gestaltung der Landschaft als Hoheitsrecht und Hoheitspflicht*, 1st ed. (Berlin: Deutsche Landesbuchhandlung, 1942), 110, 189–90, 215–16; August Wendler, *Das Problem der technischen Wetterbeeinflussung*, Probleme der kosmischen Physik 9 (Hamburg: Henri Grand, 1927).

20. Heinrich Wiepking-Jürgensmann, "Nie standen wir vor solcher Aufgabe!," *Gartenbauwirtschaft* 58, no. 19 (1941): 1; Fritz Wächtler, *Deutsches Volk, deutsche Heimat* (Munich: Deutscher Volksverlag, 1935).

21. Wiepking-Jürgensmann, "Deutsche Landschaft als deutsche Ostaufgabe," 134–35; Wiepking-Jürgensmann, *Die Landschaftsfibel*, 320ff.; see also Ursula Kellner, "Heinrich Friedrich Wiepking (1871–1973): Leben, Lehre und Werk" (PhD diss., Universität Hannover, 1998), 179; Gröning and Wolschke-Bulmahn, *Der Drang nach Osten*, 157ff.; Thomas Zeller, "'Ganz Deutschland sein Garten': Alwin Seifert und die Landschaft des Nationalsozialismus," in *Naturschutz und Nationalsozialismus*, ed. Joachim Radkau and Frank Uekötter (Frankfurt am Main: Campus Verlag, 2003), 299–300; BArch R 49/165: Wiepking-Jürgensmann, "Entwurf: Landschaftliche Richtlinien," January 16, 1942, 20. Wiepking had a penchant for designating various things as vital for the war effort. At one point, he even called "well-managed kitchens" one of the "sharpest instruments of war"; Wiepking-Jürgensmann, "Raumordnung und Landschaftsgestaltung," 18. Himmler summarized the plans for the East as the "design of a German, fortified [wehrhaft] landscape": BArch R 49/157: "Allgemeine Anordnung 7/II des Reichsführers-SS Reichskommissar für die Festigung deutschen Volkstums vom 26. November 1940, Betr. Grundsätze und Richtlinien für den ländlichen Aufbau in den Ostgebieten," n.d., 99; BArch R 4606/1605: Wiepking-Jürgensmann to Speer, February 24, 1940, n.p.

22. BArch R 49/509; see also BArch R 49/158: "Allgemeine Anordnung Nr. 20/VI/42," December 21, 1942, 55.

23. Joachim Wolschke-Bulmahn, "The Nationalization of Nature and the Naturalization of the German Nation: 'Teutonic' Trends in Early Twentieth-Century Landscape Design," in *Nature and Ideology: Natural Garden Design in the Twentieth Century*, ed. Joachim Wolschke-Bulmahn (Washington, DC: Dumbarton Oaks Research Library and Collection, 1997), 218; Erhard Mäding, *Regeln für die Gestaltung der Landschaft: Einführung in die Allgemeine Anordnung Nr. 20/VI/42* (Berlin: Deutsche Landbuchhandlung, 1943); Gröning and Wolschke-Bulmahn, *Der Drang nach Osten*, 127–34, 219.

24. BArch R 49/158: "Allgemeine Anordnung Nr. 20/VI/42," December 21, 1942; Gert Gröning, "Die 'Allgemeine Anordnung Nr. 20/VI/42'—Über die Gestaltung der Landschaft in den eingegliederten Ostgebieten," in *Der "Generalplan Ost": Hauptlinien der nationalsozialistischen Planungs- und Vernichtungspolitik*, ed. Mechtild Rössler and Sabine Schleiermacher (Berlin: Akademie Verlag, 1993), 134; see also Martin Bemmann, *Beschädigte Vegetation und sterbender Wald: Zur Entstehung eines Umweltproblems in Deutschland 1893–1970* (Göttingen: Vandenhoeck & Ruprecht, 2012), 318–26.

25. The Landscape Decree never actually passed into law; see Wolschke-Bulmahn, "The Nationalization of Nature and the Naturalization of the German Nation," 218; BArch R 73/12846: Mäding to Meyer, "Vorgang: Forschungsmittel," May 20, 1943, 16; Heiber, "Der Generalplan Ost: Vorbemerkung," 291; Hartenstein, *Neue Dorflandschaften*, 234–36.

26. Johannes Zechner, "Ewiger Wald und ewiges Volk—Der Wald als nationalsozialistischer Idealstaat," in *Naturschutz und Demokratie!?: Dokumentation der Beiträge zur Veranstaltung der*

Stiftung Naturschutzgeschichte und des Zentrums für Gartenkunst und Landschaftsarchitektur (CGL) der Leibniz Universität Hannover in Kooperation mit dem Institut für Geschichte und Theorie der Gestaltung (GTG) der Universität der Künste Berlin, ed. Gert Gröning and Joachim Wolschke-Bulmahn (Munich: Meidenbauer, 2006), 115–20; BArch R 43 II/215. For the central value of forests and wood for Nazi efforts to build an autarkic economy, see Bemmann, *Beschädigte Vegetation und sterbender Wald.* On the importance of military aims in early Nazi economic policy, see Avraham Barkai, *Nazi Economics: Ideology, Theory, and Policy* (New Haven, CT: Yale University Press, 1990), 217–24; Heinrich Rubner, *Deutsche Forstgeschichte, 1933–1945: Forstwirtschaft, Jagd, und Umwelt im NS-Staat* (St. Katharinen: Scripta Mercaturae, 1985), 91–100; Martin Bemmann, "'Wir müssen versuchen, so viel wie möglich aus dem deutschen Wald herauszuholen.' Zur ökonomischen Bedeutung des Rohstoffes Holz im 'Dritten Reich,'" *Allgemeine Forst- und Jagdzeitung* 179, no. 4 (2008): 64–69; Fehn, "'Lebensgemeinschaft von Volk und Raum,'" 222–23; Herbert Morgen, "Forstwirtschaft und Forstpolitik im neuen Osten," *Neues Bauerntum* 33 (1941): 103–7.

27. See Peter-M. Steinsiek, "Forstliche Großraumszenarien bei der Unterwerfung Osteuropas durch Hitlerdeutschland," *Vierteljahrschrift für Sozial- und Wirtschaftsgeschichte* 94, no. 2 (2007): 141–64; Joachim Wolschke-Bulmahn, "Violence as the Basis of National Socialist Landscape Planning in the 'Annexed Eastern Areas,'" in *How Green Were the Nazis? Nature, Environment, and Nation in the Third Reich,* ed. Franz-Josef Brüggemeier, Mark Cioc, and Thomas Zeller, 1st ed., Ohio University Press Series in Ecology and History (Athens: Ohio University Press, 2005), 249; BArch R 49/229: "Planungsgrundlagen für den Aufbau im Regierungsbezirk Zichenau," January 17, 1941, n.p.

28. BArch R 49/877: "Vermerk zum Vorgang: Aufforstung in den eingegliederten Ostgebieten," December 11, 1942; Barch R 49/167: Ziegler, "Vermerk über die Bereisung mit Vertretern des Reichsforstmeisters vom 22. und 23.8.1940," August 27, 1940, 5–6; see also BArch 49/2066; BArch R 3701/264: Göring to the General Government in Cracow, October 7, 1941, n.p.; BArch R 49/166. The disputes continued despite an agreement reached between the two branches of the bureaucracy in 1942; see BArch 49/169; see also Rubner, *Deutsche Forstgeschichte,* 133–40, 150–55.

29. Wasser, *Himmlers Raumplanung im Osten,* 47–59, 133–229; Hartenstein, *Neue Dorflandschaften,* 226–28; Michael G. Esch, *"Gesunde Verhältnisse": Deutsche und polnische Bevölkerungspolitik in Ostmitteleuropa 1939–1950* (Marburg: Herder-Institut, 1998), 229–52; Robert Lewis Koehl, *RKFDV: German Resettlement and Population Policy, 1939–1945: History of the Reich Commission for the Strengthening of Germandom* (Cambridge, MA: Harvard University Press, 1957), 210; David Blackbourn, *The Conquest of Nature: Water, Landscape, and the Making of Modern Germany* (New York: Norton, 2006), 306–9; BArch R 49/3144: "Allgemeine Angaben über Dorfplanung und Landschaftsgestaltung in Oberschlesien," July 12, 1943, 101; BArch R49/3513: Eppler, "Betr. Überprüfung der Gemeinde Tarnawa-Unter," September 28, 1944, n.p. Gerhard Wolf has pointed out that it is difficult to identify a unified policy of Germanization in the East, as the bureaucratic chaos and the centrifugal forces in the periphery were too strong for a clear line to emerge; see Gerhard Wolf, *Ideologie und Herrschaftsrationalität: Nationalsozialistische Germanisierungspolitik in Polen* (Hamburg: Hamburger Edition, 2012), 28–29.

30. Meyer, "Planung und Ostaufbau," 393–94; BArch R 49/165: Wiepking-Jürgensmann to Meyer, May 26, 1942, 149; BArch R 49/511: Mäding, "Vermerk betr. Generalreferat für Landschaftspflege," August 4, 1942, n.p.; BArch R 49/898: Mäding, "Sachstand und Hauptprobleme

der Landschaftsgestaltung," February 1, 1944, n.p. On the Reichsstelle für Raumordnung, see Elke Paul-Weber, "Die Reichsstelle für Raumordnung und die Ostplanung," in *Der "Generalplan Ost": Hauptlinien der nationalsozialistischen Planungs- und Vernichtungspolitik*, ed. Mechtild Rössler and Sabine Schleiermacher (Berlin: Akademie Verlag, 1993), 148–53; BArch NS 19/1739; Herbert Frank, "Dörfliche Planungen im Osten," in *Neue Dorflandschaften: Gedanken und Pläne zum ländlichen Aufbau in den neuen Ostgebieten und im Altreich*, ed. Stabshauptamt des Reichskommissars für die Festigung Deutschen Volkstums (Berlin: Sohnrey, 1943), 44–45.

31. Meyer, "Planung und Ostaufbau," 396; BArch R 49/167: "Aufforstungsflächen," June 20, 1941; Bemmann, *Beschädigte Vegetation und sterbender Wald*, 320–21; Willi Oberkrome, *"Deutsche Heimat": Nationale Konzeption und regionale Praxis von Naturschutz, Landschaftsgestaltung und Kulturpolitik in Westfalen-Lippe und Thüringen (1900–1960)* (Paderborn: Schöningh, 2004), 250–67.

32. Jürgen Zimmerer, "The Birth of the Ostland out of the Spirit of Colonialism: A Postcolonial Perspective on the Nazi Policy of Conquest and Extermination," *Patterns of Prejudice* 39, no. 2 (2005): 198.

33. BArch R 49/2388: Mäding, "Richtlinien zur Landschaftsgestaltung," n.d., 7; BArch R 49/2502: Abteilung Raumuntersuchung, "Kurzer Überblick über Osteuropa," n.d., n.p.; H. Kuron, "Die Bodenerosion in Europa: Eine zusammenfassende Darstellung ohne die Gebirge," *Der Forschungsdienst* 16, no. 1 (1943): 6–20.

34. Hermann Leiter, *Ukraine: Der Südosten Europas in der Wirtschaft Großdeutschlands*, ed. NSDAP Gau Wien (Vienna: Lang & Gratzenberger, 1942), 7.

35. Joachim Radkau, *Technik in Deutschland. Vom 18. Jahrhundert bis heute* (Frankfurt am Main: Campus, 2008), 301. On the penchant for thinking on the large scale among Nazi planners, see Trevor J. Barnes and Claudio Minca, "Nazi Spatial Theory: The Dark Geographies of Carl Schmitt and Walter Christaller," *Annals of the Association of American Geographers* 103, no. 3 (May 1, 2013): 669–87; see also Thomas Lekan, "'It Shall Be the Whole Landscape': The Reich Nature Protection Law and Regional Planning in the Third Reich," in *How Green Were the Nazis? Nature, Environment, and Nation in the Third Reich*, ed. Franz-Josef Brüggemeier, Mark Cioc, and Thomas Zeller (Athens: Ohio University Press, 2005), 73–100; BArch R 49/509; Zeller, *Driving Germany*, 67.

36. See Mechtild Rössler, "Konrad Meyer und der 'Generalplan Ost' in der Beurteilung der Nürnberger Prozesse," in *Der "Generalplan Ost": Hauptlinien der nationalsozialistischen Planungs- und Vernichtungspolitik*, ed. Mechtild Rössler and Sabine Schleiermacher (Berlin: Akademie Verlag, 1993), 356–65; Gert Gröning, "Teutonic Myth, Rubble, and Recovery: Landscape Architecture in Germany," in *The Architecture of Landscape, 1940–1960*, ed. Marc Treib (Philadelphia: University of Pennsylvania Press, 2002), 120–54. Wiepking's statement at the Nuremberg Trials is cited in Gröning and Wolschke-Bulmahn, *Der Drang nach Osten*, 219; ADM NL 133/008: Seifert, "Vom Sinn der vielen Dürre," 1948, 6.

37. Lutz Mackensen, ed., *Deutsche Heimat ohne Deutsche: Ein ostdeutsches Heimatbuch* (Braunschweig: G. Westermann, 1951). On the politics of memory and the continuing significance of the "German East" in the postwar period, see Andrew Demshuk, *The Lost German East: Forced Migration and the Politics of Memory, 1945–1970* (New York: Cambridge University Press, 2012); see also David Blackbourn, "'The Garden of Our Hearts': Landscape, Nature, and Local Identity in the German East," in *Localism, Landscape, and the Ambiguities of Place:*

German-Speaking Central Europe, 1860–1930, ed. James N. Retallack and David Blackbourn (Toronto: University of Toronto Press, 2007), 158–59.

38. Anton Olbrich, *Windschutzpflanzungen* (Hannover: M. & H. Schaper, 1949); see also Gröning and Wolschke-Bulmahn, *Der Drang nach Osten*, 122–23; Stephen Brain, "The Great Stalin Plan for the Transformation of Nature," *Environmental History* 15, no. 4 (October 1, 2010): 670–700; Hans Findeisen, *Nach Südrussland greift die Wüste* (Augsburg: Institut für Menschen- und Menschheitskunde, 1950), 1.

Chapter 8

1. For a concise story of climate change research in the second half of the twentieth century, see Spencer Weart, *The Discovery of Global Warming* (Cambridge, MA: Harvard University Press, 2003); Matthias Heymann and Dania Achermann, "From Climatology to Climate Science in the Twentieth Century," in *The Palgrave Handbook of Climate History*, ed. Sam White, Christian Pfister, and Franz Mauelshagen (London: Palgrave Macmillan, 2018), 605–32. For the history of atmospheric weather and climate engineering (with a focus on the second half of the twentieth century), see James Rodger Fleming, *Fixing the Sky: The Checkered History of Weather and Climate Control* (New York: Columbia University Press, 2010); Noah Byron Bonnheim, "History of Climate Engineering," *Wiley Interdisciplinary Reviews: Climate Change* 1, no. 6 (November 1, 2010): 891–97; Jacob Darwin Hamblin, *Arming Mother Nature: The Birth of Catastrophic Environmentalism* (New York: Oxford University Press, 2013), 108–28. For an overview of the globalization of desertification concerns and research over the twentieth century, see Tor A. Benjaminsen and Pierre Hiernaux, "From Desiccation to Global Climate Change: A History of the Desertification Narrative in the West African Sahel, 1900–2018," *Global Environment* 12, no. 1 (March 2019): 206–36.

2. Deborah R. Coen, *Climate in Motion: Science, Empire, and the Problem of Scale* (Chicago: University of Chicago Press, 2018); Matthias Heymann, "Klimakonstruktionen," *NTM Zeitschrift für Geschichte der Wissenschaften, Technik und Medizin* 17, no. 2 (May 1, 2009): 171–97; Matthias Heymann, "The Evolution of Climate Ideas and Knowledge," *Wiley Interdisciplinary Reviews: Climate Change* 1, no. 4 (2010): 581–97; Sverker Sörlin, "The Global Warming That Did Not Happen: Historicizing Glaciology and Climate Change," in *Nature's End: History and the Environment*, ed. Paul Warde and Sverker Sörlin (Houndmills, Basingstoke: Palgrave Macmillan, 2009), 93–114.

3. Sverker Sörlin tells a parallel—and indeed overlapping—history of the division between glaciology and meteorology and the development of "distinct epistemic communities" in "Narratives and Counter-Narratives of Climate Change: North Atlantic Glaciology and Meteorology, c. 1930–1955," *Journal of Historical Geography* 35, no. 2 (April 2009): 237–55.

4. Sharon E. Nicholson, "Revised Rainfall Series for the West African Subtropics," *Monthly Weather Review* 107, no. 5 (May 1979): 620–23; Mike Hulme, "Climatic Perspectives on Sahelian Desiccation: 1973–1998," *Global Environmental Change* 11, no. 1 (April 2001): 19–29; S. E. Nicholson, "Climatic Variations in the Sahel and Other African Regions during the Past Five Centuries," *Journal of Arid Environments* 1 (1978): 3–24; Yen-Ting Hwang, Dargan M. W. Frierson, and Sarah M. Kang, "Anthropogenic Sulfate Aerosol and the Southward Shift of Tropical Precipitation in the Late 20th Century," *Geophysical Research Letters* 40, no. 11 (June 16, 2013): 2845;

Nicholas Wade, "Sahelian Drought: No Victory for Western Aid," *Science* 185, no. 4147 (July 19, 1974): 234–37.

5. Marybeth Long Martello, "Expert Advice and Desertification Policy: Past Experience and Current Challenges," *Global Environmental Politics* 4, no. 3 (2004): 94–95; A. Aubréville, *Climats, forêts et désertification de l'Afrique tropicale* (Paris: Société d'éditions géographiques, maritimes et coloniales, 1949); see also Jeremy Swift, "Desertification: Narratives, Winners and Losers," in *The Lie of the Land: Challenging Received Wisdom on the African Environment*, ed. Melissa Leach and Robin Mearns (London: International African Institute, 1996), 73–77.

6. Joseph Otterman, "Baring High-Albedo Soils by Overgrazing: A Hypothesized Desertification Mechanism," *Science* 186, no. 4163 (November 8, 1974): 531–33. For a short historical overview of desertification research in Africa, see James C. McCann, "Climate and Causation in African History," *International Journal of African Historical Studies* 32, no. 2/3 (January 1, 1999): 270–73; J. G. Charney, "Dynamics of Deserts and Drought in the Sahel," *Quarterly Journal of the Royal Meteorological Society* 101, no. 428 (1975): 193–202. For a critique of Charney's findings, see S. B. Idso, "A Note on Some Recently Proposed Mechanisms of Genesis of Deserts," *Quarterly Journal of the Royal Meteorological Society* 103, no. 436 (1977): 369–70; H. H. Lamb, "Some Comments on the Drought in Recent Years in the Sahel-Ethiopian Zone of North Africa," *African Environment. Special Report*, no. 6 (1977): 33–37.

7. For a historical overview of desertification policies, see M. Kassas, "Desertification: A General Review," *Journal of Arid Environments* 30, no. 2 (June 1995): 115–28. In 1961, UNESCO, in collaboration with the World Meteorological Organization, organized a symposium in Rome on climate changes. The published proceedings of the meeting became an important collection on the state of climatological knowledge at the time, reflecting the diversity of approaches in the field. See World Meteorological Organization and UNESCO Arid Zone Programme, eds., *Changes of Climate: Proceedings of the Rome Symposium Organized by UNESCO and the World Meteorological Organization* (Paris: UNESCO, 1963); United Nations, *United Nations Conference on Desertification, 29 August—9 September 1977: Round-Up, Plan of Action, and Resolutions* (New York: United Nations, 1978), 4.

8. United Nations, *Conference on Desertification*, 5.

9. The most far-reaching part of the program was a comprehensive plan of centrally coordinated local, national, and international action to arrest and reverse desertification in Africa and beyond; see United Nations, *Desertification: Its Causes and Consequences* (Oxford: Pergamon Press, 1977), 44–60; United Nations, *Conference on Desertification*.

10. Swift, "Desertification: Narratives, Winners and Losers," 73–90. A widely cited 1990 study listed four causes of desertification, namely "overcultivation, overgrazing, poor irrigation management, and deforestation," all of which were described as the direct effect of local human action; see Alan Grainger, *The Threatening Desert: Controlling Desertification* (London: Earthscan, 1990), 65–106; United Nations, *Report of the United Nations Conference on Environment and Development, Rio de Janeiro, 3–14 June 1992*, 3 vols. (New York: United Nations, 1993); United Nations Environment Programme, *Status of Desertification and Implementation of the United Nations Plan of Action to Combat Desertification: Report of the Executive Director* (Nairobi, Kenya: United Nations Environmental Programme, 1991). See also Mike Hulme and Mick Kelly, "Exploring the Links between Desertification and Climate Change," *Environment: Science and Policy*

for Sustainable Development 35, no. 6 (1993): 5–6; United Nations Environment Programme, *World Atlas of Desertification* (Sevenoaks: Edward Arnold, 1992).

11. United Nations, *United Nations Convention to Combat Desertification in Those Countries Experiencing Serious Drought and/or Desertification, Particularly in Africa* (Geneva: Interim Secretariat for the Convention to Combat Desertification, 1997); see also Elisabeth Corell, *The Negotiable Desert: Expert Knowledge in the Negotiations of the Convention to Combat Desertification* (Linköping: Department of Water and Environmental Studies, University of Linköping, 1999); W. Neil Adger et al., "Advancing a Political Ecology of Global Environmental Discourses," *Development and Change* 32, no. 4 (2001): 681–715; S. M. Herrmann and C. F. Hutchinson, "The Changing Contexts of the Desertification Debate," *Journal of Arid Environments* 63, no. 3 (November 2005): 551.

12. Mike Hulme, R. Marsh, and P. D. Jones, "Global Changes in a Humidity Index Between 1931–60 and 1961–90," *Climate Research* 2, no. 1 (1992): 1–22; Melissa Leach and Robin Mearns, "Challenging Received Wisdom in Africa," in *The Lie of the Land: Challenging Received Wisdom on the African Environment*, ed. Melissa Leach and Robin Mearns (London: International African Institute, 1996), 1–33; Michael Mortimore, *Roots in the African Dust: Sustaining the Sub-Saharan Drylands* (Cambridge: Cambridge University Press, 1998), 22–25; Compton J. Tucker, Harold E. Dregne, and Wilbur W. Newcomb, "Expansion and Contraction of the Sahara Desert from 1980 to 1990," *Science* 253, no. 5017 (July 19, 1991): 299–301; R. Monastersky, "Satellites Expose Myth of Marching Sahara," *Science News* 140, no. 3 (July 20, 1991): 38; Compton J. Tucker and Sharon E. Nicholson, "Variations in the Size of the Sahara Desert from 1980 to 1997," *Ambio* 28, no. 7 (November 1, 1999): 587–91.

13. C. K. Folland, T. N. Palmer, and D. E. Parker, "Sahel Rainfall and Worldwide Sea Temperatures, 1901–85," *Nature* 320, no. 6063 (1986): 602–7. For an attempt to construct a predictive model of precipitation levels on the basis of SST variations, see Chris Folland et al., "Prediction of Seasonal Rainfall in the Sahel Region Using Empirical and Dynamical Methods," *Journal of Forecasting* 10, nos. 1–2 (1991): 21–56; Hulme and Kelly, "Exploring the Links between Desertification and Climate Change," 41; Ulf Helldén, "Desertification: Time for an Assessment?," *Ambio* 20, no. 8 (December 1, 1991): 383.

14. Lennart Olsson, "Desertification in Africa—A Critique and an Alternative Approach," *GeoJournal* 31, no. 1 (September 1, 1993): 23–31; D. S. G. Thomas and N. J. Middleton, *Desertification: Exploding the Myth* (Chichester: Wiley, 1994). The adverse effect of political and bureaucratic entanglements on research by the US Forest Service has been described in Ashley L. Schiff, *Fire and Water: Scientific Heresy in the Forest Service* (Cambridge, MA: Harvard University Press, 1962); Vasant K. Saberwal, "Science and the Desiccationist Discourse of the 20th Century," *Environment and History* 4, no. 3 (October 1, 1998): 309–43.

15. James Fairhead and Melissa Leach, *Misreading the African Landscape: Society and Ecology in a Forest-Savanna Mosaic* (Cambridge: Cambridge University Press, 1996). For a discussion on the applicability of equilibrium models to (African) drylands, see Mortimore, *Roots in the African Dust*; James Fairhead and Melissa Leach, *Reframing Deforestation—Global Analyses and Local Realities: Studies in West Africa* (London: Routledge, 1998); Herrmann and Hutchinson, "The Changing Contexts of the Desertification Debate"; J. V. Vogt et al., "Monitoring and Assessment of Land Degradation and Desertification: Towards New Conceptual and Integrated

Approaches," *Land Degradation & Development* 22, no. 2 (March 1, 2011): 150–65; Elina Andersson, Sara Brogaard, and Lennart Olsson, "The Political Ecology of Land Degradation," *Annual Review of Environment and Resources* 36, no. 1 (2011): 295–319. See also Roy Behnke and Michael Mortimore, eds., *The End of Desertification? Disputing Environmental Change in the Drylands,* Springer Earth System Sciences (Berlin: Springer, 2016).

16. For early intimations of the connection, see Fred Pearce, "A Sea Change in the Sahel," *New Scientist* 129, no. 1754 (1991): 31; Hulme and Kelly, "Exploring the Links Between Desertification and Climate Change"; Hulme, "Climatic Perspectives," 25–26.

17. Yongkang Xue, "Biosphere Feedback on Regional Climate in Tropical North Africa," *Quarterly Journal of the Royal Meteorological Society* 123, no. 542 (July 1, 1997): 1483–1515; R. C. Balling, "Impact of Desertification on Regional and Global Warming," *Bulletin of the American Meteorological Society* 72, no. 2 (1991): 232–34; Ye Wang and Xiaodong Yan, "Climate Change Induced by Southern Hemisphere Desertification," *Physics and Chemistry of the Earth, Parts A/B/C* 102 (December 1, 2017): 40–47; Max Rietkerk et al., "Local Ecosystem Feedbacks and Critical Transitions in the Climate," *Ecological Complexity* 8, no. 3 (September 2011): 223–28; Pierre Marc Johnson, Karel Mayrand, and Marc Paquin, eds., *Governing Global Desertification— Linking Environmental Degradation, Poverty and Participation* (London: Routledge, 2016).

18. David P. Rowell et al., "Variability of Summer Rainfall Over Tropical North Africa (1906–92): Observations and Modelling," *Quarterly Journal of the Royal Meteorological Society* 121, no. 523 (April 1, 1995): 669–704; Rattan Lal, "Climate Change and Soil Degradation Mitigation by Sustainable Management of Soils and Other Natural Resources," *Agricultural Research* 1, no. 3 (September 1, 2012): 199–212. For an overview, see Alessandra Giannini, Michela Biasutti, and Michel M. Verstraete, "A Climate Model-Based Review of Drought in the Sahel: Desertification, the Re-greening and Climate Change," *Global and Planetary Change* 64, no. 3–4 (December 2008): 119–28; J. Huang et al., "Dryland Climate Change: Recent Progress and Challenges," *Reviews of Geophysics* 55, no. 3 (September 1, 2017): 719–78.

19. J. F. Reynolds and D. M. Stafford Smith, "Desertification: A New Paradigm for an Old Problem," in *Global Desertification: Do Humans Cause Deserts?*, ed. J. F. Reynolds and D. M. Stafford Smith (Berlin: Dahlem University Press, 2002), 403–24.

20. R. S. Deese, "The Artifact of Nature: 'Spaceship Earth' and the Dawn of Global Environmentalism," *Endeavour* 33, no. 2 (June 2009): 70–75; Sheila Jasanoff and Marybeth Long Martello, "Heaven and Earth: The Politics of Environmental Images," in *Earthly Politics: Local and Global in Environmental Governance* (Cambridge, MA: MIT Press, 2004), 31–54; Sabine Höhler, *Spaceship Earth in the Environmental Age, 1960–1990* (Abingdon: Routledge, 2016).

21. See Milutin Milanković, *Canon of Insolation and the Ice-Age Problem (Kanon Der Erdbestrahlung Und Seine Anwendung Auf Das Eiszeitenproblem)* (Jerusalem: Israel Program for Scientific Translations, 1969). See also John Imbrie and Katherine Palmer Imbrie, *Ice Ages: Solving the Mystery* (Cambridge, MA: Harvard University Press, 1986), 141–46; Richard A. Kerr, "Milankovitch Climate Cycles through the Ages," *Science* 235, no. 4792 (1987): 973–74; David A. Hodell, "The Smoking Gun of the Ice Ages," *Science* 354, no. 6317 (December 9, 2016): 1235–36. For one of the most highly influential papers on a global cooling trend, see J. M. Mitchell, "On the World-Wide Pattern of Secular Temperature Change," in *Changes of Climate: Proceedings of the Rome Symposium Organized by UNESCO and the World Meteorological Organization* (Paris: UNESCO, 1963), 161–81; Lowell Ponte, *The Cooling: Has the Next Ice Age Already Begun?*

(Englewood Cliffs, NJ: Prentice Hall, 1976); Howard A. Wilcox, *Hothouse Earth* (New York: Praeger, 1975); Thomas C. Peterson, William M. Connolley, and John Fleck, "The Myth of the 1970s Global Cooling Scientific Consensus," *Bulletin of the American Meteorological Society* 89, no. 9 (September 2008): 1325–37.

22. Roger A. Pielke and William R. Cotton, *Human Impacts on Weather and Climate*, 2nd ed. (Cambridge: Cambridge University Press, 2007), 1–71; Fleming, *Fixing the Sky*, 165–88; Georg Breuer, *Weather Modification: Prospects and Problems* (Cambridge: Cambridge University Press, 1980), 144–57. On the early history of cloud seeding from the late nineteenth century, see Clark C Spence, *The Rainmakers: American "Pluviculture" to World War II* (Lincoln: University of Nebraska Press, 1980), 117–24; Kristine Harper, *Weather by the Numbers: The Genesis of Modern Meteorology* (Cambridge, MA: MIT Press, 2008), 4–5, 91–119; National Research Council, Assembly of Mathematical and Physical Sciences, Committee on Atmospheric Sciences, *The Atmospheric Sciences: Problems and Applications* (Washington, DC: National Academy of Sciences, 1977), 86–98; Edith Brown Weiss, "International Responses to Weather Modification," *International Organization* 29, no. 3 (July 1, 1975): 805.

23. On French conceptions of the Sahara and visions for its exploitation in the 1950s and 1960s, see George R. Trumbull, "Body of Work: Water and the Reimagining of the Sahara in the Era of Decolonization," in *Environmental Imaginaries of the Middle East and North Africa*, ed. Diana K. Davis and Edmund Burke (Athens: Ohio University Press, 2011), 87–112; Pierre Cornet, *Sahara, terre de demain* (Paris: Nouvelles éditions latines, 1957); Daniel Strasser, "L'organisation économique du Sahara," *Comptes rendus mensuels des séances de l'Académie des Sciences d'Outre-Mer* 17 (July 1957): 232–44.

24. Much of the information in this and the following paragraphs is taken from René Létolle and Hocine Bendjoudi, *Histoires d'une mer au Sahara: utopies et politiques* (Paris: Harmattan, 1997), 117–27; Raymond Furon, *The Problem of Water: A World Study* (London: Faber, 1967), 157; R. Bonhours, "La mer en pénétrant dans le Sahara poduira de l'énergie électrique et fertilisera une partie du désert," *Science et Vie*, September 1954. See also Jean-Robert Henry, Jean-Louis Marçot, and Jean-Yves Moisseron, "Développer le désert: Anciennes et nouvelles utopies," *L'Année du Maghreb*, no. 7 (September 1, 2011): 115–47. For Kevran's work on low-energy transmutations of one element into another or the "Kevran effect," see C. Louis Kervran, *Transmutations biologiques: Métabolismes aberrants de l'azote, le potassium et le magnésium* (Paris: Librairie Maloine, 1962); F. Charles-Roux and Jean Goby, "Ferdinand de Lesseps et le projet de mer intérieure africaine," *Revue des deux Mondes* 15 (1957): 385–404; Georg Gerster, *Sahara: Desert of Destiny* (New York: Coward-McCann, 1961), 270; Kurt Hiehle, "L'Irrigation du Sahara," *Industries et travaux d'outre-mer: Afrique, Amérique Latine, Asie*, July 1955, 415–20.

25. On France's nuclear program and its important cultural and political role in postwar France, see Gabrielle Hecht, *The Radiance of France: Nuclear Power and National Identity after World War II* (Cambridge, MA: MIT Press, 1998); Paul Denarié, "Une mer intérieure au Sahara," *Sahara de demain: revue d'informations mensuelle sur tous les problèmes sahariens* 5 (1958): 16–19; Létolle and Bendjoudi, *Histoires d'une mer au Sahara*, 118–19.

26. Martin Walker, "Drought: Nature and Well-Meaning Men Have Combined to Produce a Catastrophe Imperiling Many Millions," *New York Times Magazine*, June 9, 1974.

27. Kevin Lowther and C. Payne Lucas, "A Plan to Make the Sahara Bloom," *Washington Post*, August 4, 1974; G. Ali Heshmati and Victor R. Squires, eds., *Combating Desertification in Asia,*

Africa and the Middle East: Proven Practices (Dordrecht: Springer, 2013), 50–51, 187; Elius Levin, "Growing China's Great Green Wall," *ECOS* 2005, no. 127 (November 21, 2005): 13. On some recent criticism of the Chinese Green Wall, see X. M. Wang et al., "Has the Three Norths Forest Shelterbelt Program Solved the Desertification and Dust Storm Problems in Arid and Semiarid China?," *Journal of Arid Environments* 74, no. 1 (January 2010): 13–22; Shixiong Cao et al., "Excessive Reliance on Afforestation in China's Arid and Semi-arid Regions: Lessons in Ecological Restoration," *Earth-Science Reviews* 104, no. 4 (February 2011): 240–45; Hong Jiang, "Taking Down the 'Great Green Wall': The Science and Policy Discourse of Desertification and Its Control in China," in *The End of Desertification?: Disputing Environmental Change in the Drylands*, ed. Roy Behnke and Michael Mortimore, Springer Earth System Sciences (Berlin: Springer, 2016), 513–36. For historical contextualization, see Mark Elvin, *The Retreat of the Elephants: An Environmental History of China* (New Haven, CT: Yale University Press, 2004), 19–39. The quote from the USAID worker is from Burkhard Bilger, "The Great Oasis," *New Yorker*, December 19, 2011.

28. P. Rognon, *Biographie d'un désert* (Paris: Plon, 1989), 336–37; V. Badescu, Richard B. Cathcart, and A. A. Bolonkin, "Sand Dune Fixation: A Solar-Powered Sahara Seawater Pipeline Macroproject," *Land Degradation & Development* 19, no. 6 (2008): 676–91.

29. Létolle and Bendjoudi, *Histoires d'une mer au Sahara*, 120–26.

30. Richard Cathcart, *Herman Sörgel* (Monticello, IL: Vance Bibliographies, 1980); Nicola M. Pugno, Richard B. Cathcart, and Joseph J. Friedlander, "Treeing the CATS: Artificial Gulf Formation by the Chotts Algeria–Tunisia Scheme," in *Macro-Engineering Seawater in Unique Environments*, ed. Viorel Badescu et al., Environmental Science and Engineering (Berlin: Springer, 2011), 489–517.

31. Ari Daniel, "What's a Lake Doing in the Middle of the Desert?," *All Things Considered* (National Public Radio, October 26, 2012), http://www.npr.org/2012/10/26/163723606/whats-a-lake-doing-in-the-middle-of-the-desert.

32. See, e.g., Lambert K. Smedema and Karim Shiati, "Irrigation and Salinity: A Perspective Review of the Salinity Hazards of Irrigation Development in the Arid Zone," *Irrigation and Drainage Systems* 16, no. 2 (May 1, 2002): 161–74.

33. National Academy of Sciences, Committee on Science, Engineering, and Public Policy, *Policy Implications of Greenhouse Warming: Mitigation, Adaptation, and the Science Base* (Washington, DC: National Academy Press, 1992).

34. Michael Specter, "The Climate Fixers," *New Yorker*, April 14, 2012. For both sympathetic and critical reviews or recent geoengineering ideas, see David W. Keith, Gernot Wagner, and Claire L. Zabel, "Solar Geoengineering Reduces Atmospheric Carbon Burden," *Nature Climate Change* 7 (September 1, 2017): 617–19; Naomi E. Vaughan and Timothy M. Lenton, "A Review of Climate Geoengineering Proposals," *Climatic Change* 109, nos. 3–4 (December 2011): 745–90; Michael Brzoska, P. Michael Link, and Götz Neuneck, "Geoengineering—Möglichkeiten und Risiken," *S+F Sicherheit und Frieden* 30, no. 4 (2012); Elizabeth T. Burns et al., "What Do People Think When They Think about Solar Geoengineering? A Review of Empirical Social Science Literature, and Prospects for Future Research," *Earth's Future* 4, no. 11 (November 1, 2016): 536–42.

35. IPCC, "Summary for Policymakers," in *Climate Change 2013: The Physical Science Basis. Contribution of Working Group I to the Fifth Assessment Report of the Intergovernmental Panel on Climate Change*, ed. T. F. Stocker et al. (Cambridge: Cambridge University Press, 2013), 27. The

Hugh Huhnt quote is taken from Specter, "The Climate Fixers." The SPICE project aims to study the climatic effects of the injection of the stratosphere with sulfur aerosols.

36. See Mike Hulme, "Reducing the Future to Climate: A Story of Climate Determinism and Reductionism," *Osiris* 26, no. 1 (February 2011): 245–66; Amy Dahan, "Putting the Earth System in a Numerical Box? The Evolution from Climate Modeling toward Global Change," *Studies in History and Philosophy of Science Part B: Studies in History and Philosophy of Modern Physics* 41, no. 3 (September 2010): 282–92; Matthias Heymann, "Understanding and Misunderstanding Computer Simulation: The Case of Atmospheric and Climate Science—An Introduction," *Studies in History and Philosophy of Science Part B: Studies in History and Philosophy of Modern Physics* 41, no. 3 (September 2010): 193–200. The problems of heterogeneous datasets and interdisciplinary communication could also be described as "science friction"; see Paul N. Edwards et al., "Science Friction: Data, Metadata, and Collaboration," *Social Studies of Science* 41, no. 5 (October 1, 2011): 667–90.

37. James Rodger Fleming, "Will Geo-Engineering Bring Security and Peace? What Does History Tell Us?," *S+F Sicherheit Und Frieden* 30, no. 4 (2012): 11; "Historical Perspectives on 'Fixing the Sky': Statement of Dr. James Fleming, Professor and Director of Science, Technology and Society, Colby College Before the Committee on Science and Technology, U.S. House of Representatives" (Washington, DC, November 5, 2009), 7, https://science.house.gov/hearings /geoengineering-assessing-the-implications-of-large-scale-climate-intervention.

38. Mike Hulme, "The Conquering of Climate: Discourses of Fear and Their Dissolution," *Geographical Journal* 174, no. 1 (March 1, 2008): 5–16; Naomi Oreskes, "The Scientific Consensus on Climate Change," *Science* 306, no. 5702 (December 3, 2004): 1686; Naomi Oreskes and Erik M. Conway, *Merchants of Doubt: How a Handful of Scientists Obscured the Truth on Issues from Tobacco Smoke to Global Warming* (New York: Bloomsbury, 2011).

39. See, e.g., Roger A. Pielke, "Land Use and Climate Change," *Science* 310, no. 5754 (December 9, 2005): 1625–26; Pielke and Cotton, *Human Impacts on Weather and Climate*, 102–50; Mike Hulme, "Geographical Work at the Boundaries of Climate Change," *Transactions of the Institute of British Geographers* 33, no. 1 (January 2008): 5–11; Amy Dahan-Dalmedico, "Climate Expertise: Between Scientific Credibility and Geopolitical Imperatives," *Interdisciplinary Science Reviews* 33, no. 1 (March 2008): 72.

ARCHIVES

AAS   Archives de l'Académie des Sciences, Paris, France

ADM   Archiv des Deutschen Museums, Munich, Germany

AMAE   Archives du Ministère des Affaires Étrangères, La Courneuve, France

ANOM   Archives Nationales d'Outre Mer, Aix-en-Provence, France

BARCH   Bundesarchiv, Koblenz and Berlin, Germany

GSTAPK   Geheimes Staatsarchiv Preußischer Kulturbesitz, Berlin, Germany

HLHU   Houghton Library, Harvard University, Cambridge, MA

NA   The National Archives, Kew, United Kingdom

NLSO   Niedersächsisches Landesarchiv & Staatsarchiv, Osnabrück, Germany

INDEX

Page numbers in *italics* refer to figures and tables.

A NOTE ON THE TYPE

This book has been composed in Arno, an Old-style serif typeface in the
classic Venetian tradition, designed by Robert Slimbach at Adobe.

GPSR Authorized Representative: Easy Access System Europe - Mustamäe tee
50, 10621 Tallinn, Estonia, gpsr.requests@easproject.com

www.ingramcontent.com/pod-product-compliance
Ingram Content Group UK Ltd.
Pitfield, Milton Keynes, MK11 3LW, UK
UKHW040205190325
456107UK00005B/133